Mitigating Global Climate Change - Enhancing Adaptation, Evaluation, and Restoration of Mountain Ecosystems

Edited by Ling Zhang, Shuli Wang and Liangying Liu

Published in London, United Kingdom

Mitigating Global Climate Change - Enhancing Adaptation, Evaluation, and Restoration of Mountain Ecosystems
http://dx.doi.org/10.5772/intechopen.111011
Edited by Ling Zhang, Shuli Wang and Liangying Liu

Assistant to the Editors: Nasir Shad

Contributors
Abubakr Osman, Ahmed A. H. Siddig, Ahmed Hasoba, Alan Haney, Alena Anatoljevna Isupova, Czimber Kornel, Deepak K. Gautam, Eiman E. Deiab, Emad Yasin, Him Lal Shrestha, John Asatryan, Ling Zhang, Mohamed Hemida, Ngamindra R. Dahal, Qifei Cai, Roshan M. Man Bajracharya, Steven I. Apfelbaum, Xiaodong Geng, Xiaomin Guo, Yanmei Wang, Zhen Liu, Zhi Li

First published in London, United Kingdom, 2024 by IntechOpen
IntechOpen is the global imprint of INTECHOPEN LIMITED, registered in England and Wales, registration number: 11086078, 167-169 Great Portland Street, London, W1W 5PF, United Kingdom

British Library Cataloguing-in-Publication Data
A catalogue record for this book is available from the British Library

Additional hard and PDF copies can be obtained from orders@intechopen.com

Mitigating Global Climate Change - Enhancing Adaptation, Evaluation, and Restoration of Mountain Ecosystems
Edited by Ling Zhang, Shuli Wang and Liangying Liu
p. cm.

This title is part of the Environmental Sciences Book Series, Volume 18
Topic: Ecosystems and Biodiversity
Series Editor: J. Kevin Summers
Topic Editors: Salustiano Mato, Josefina Garrido and Francisco Ramil

Print ISBN 978-0-85014-051-4
Online ISBN 978-0-85014-052-1
eBook (PDF) ISBN 978-0-85014-053-8
ISSN 2754-6713

For EU product safety concerns:
IN TECH d.o.o., Prolaz Marije Krucifikse Kozulić 3, 51000 Rijeka, Croatia, info@intechopen.com or visit our website at intechopen.com.

IntechOpen Book Series
Environmental Sciences
Volume 18

Aims and Scope of the Series

Scientists have long researched to understand the environment and man's place in it. The search for this knowledge grows in importance as rapid increases in population and economic development intensify humans' stresses on ecosystems. Fortunately, rapid increases in multiple scientific areas are advancing our understanding of environmental sciences. Breakthroughs in computing, molecular biology, ecology, and sustainability science are enhancing our ability to utilize environmental sciences to address real-world problems.

The four topics of this book series - Pollution; Environmental Resilience and Management; Ecosystems and Biodiversity; and Water Science - will address important areas of advancement in the environmental sciences. They will represent an excellent initial grouping of published works on these critical topics.

Meet the Series Editor

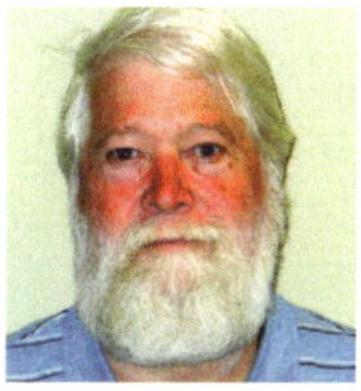

J. Kevin Summers is a Senior Research Ecologist at the Environmental Protection Agency's (EPA) Gulf Ecosystem Measurement and Modeling Division. He is currently working with colleagues in the Sustainable and Healthy Communities Program to develop an index of community resilience to natural hazards, an index of human well-being that can be linked to changes in the ecosystem, social and economic services, and a community sustainability tool for communities with populations under 40,000. He leads research efforts for indicator and indices development. Dr. Summers is a systems ecologist and began his career at the EPA in 1989 and has worked in various programs and capacities. This includes leading the National Coastal Assessment in collaboration with the Office of Water which culminated in the award-winning National Coastal Condition Report series (four volumes between 2001 and 2012), and which integrates water quality, sediment quality, habitat, and biological data to assess the ecosystem condition of the United States estuaries. He was acting National Program Director for Ecology for the EPA between 2004 and 2006. He has authored approximately 150 peer-reviewed journal articles, book chapters, and reports and has received many awards for technical accomplishments from the EPA and from outside of the agency. Dr. Summers holds a BA in Zoology and Psychology, an MA in Ecology, and Ph.D. in Systems Ecology/Biology.

Meet the Volume Editors

Dr. Ling Zhang obtained his Ph.D. from Nanjing Agricultural University, China. He visited the Department of Ecology and Evolutionary Biology, Rice University, USA, from 2012 to 2013. He is currently working as a professor in the College of Forestry, Jiangxi Agricultural University, China, supervising graduate students. Dr. Zhang studies global change biology, forest ecology, and soil carbon and nitrogen cycling. He has published more than 90 papers related to global change ecology or forest ecology, and has authored or co-authored seven books in recent years. He also serves as editor of ISI journals, such as *Journal of Plant Ecology*, *Journal of Integrative Agriculture*, *Plant Ecology*, and *Frontiers in Forests and Climate Change* and as a reviewer for more than 40 ISI journals.

Dr. Shuli Wang obtained her Ph.D. in soil and water conservation and desertification control from Jiangxi Agricultural University, China. She is currently working as a lecturer at the College of Forestry, Jiangxi Agricultural University, China. Dr. Wang studies global change biology, forest ecology, and soil carbon and nitrogen cycling. She has published more than 20 papers related to global change ecology or forest ecology.

Dr. Liangying Liu obtained her Ph.D. in forest cultivation from Jiangxi Agricultural University, China. She is currently working as an experimentalist at College of Forestry, Jiangxi Agricultural University, China. Dr. Liu studies forest ecology and soil carbon and nitrogen cycling. She has published more than eight papers and obtained four science and technology awards related to forest ecology.

Contents

Preface

Global climate change poses a significant threat to ecosystem stability, biodiversity, and function worldwide. Mountain ecosystems, while providing essential resources such as materials, food, energy, and recreational opportunities, are particularly vulnerable to degradation, erosion, and biodiversity loss due to exploitation. These threats not only impact local ecosystems but also contribute to global environmental changes.

Despite their susceptibility to environmental shifts, mountain ecosystems play crucial roles in preserving biological diversity and mitigating the effects of global climate change. Therefore, it is imperative to leverage the latest technologies, methodologies, and strategies to enhance our understanding of adaptation, evaluation, and restoration efforts in mountainous regions.

Agriculture, forests, meadows, and abandoned tailings or quarries within mountain ecosystems are increasingly recognized for their importance in adaptation, evaluation, and restoration practices. These elements are closely linked to the stability of mountain ecosystems and merit special attention in conservation efforts.

This volume introduces cutting-edge approaches to the adaptation, evaluation, and restoration of mountain ecosystems. By emphasizing sustainability and resilience, these efforts aim to safeguard mountain ecosystems and the communities that depend on them, ultimately promoting the well-being of human society.

We gratefully acknowledge the valuable assistance provided by Dr. Nasir Shad, who is now studying as a Ph.D. student at Nanjing University of Information Science and Technology, during the editorial process. We appreciate his dedication and effort in reviewing and editing several submitted chapters. Additionally, we extend our thanks to the editing office for their diligent work, including communication and other tasks associated with the publication of all chapters.

Ling Zhang, Shuli Wang and Liangying Liu
College of Forestry,
Jiangxi Agricultural University,
Nanchang, China

Section 1

Adaptation of Mountain Ecosystems to Global Climate Change

Chapter 1

Diversified Agroforestry for Climate Change Adaptation and Mitigation in the Himalayan Region: Potential for Achieving Multiple Benefits

Roshan M. Bajracharya, Deepak K. Gautam, Ngamindra R. Dahal and Him Lal Shrestha

Abstract

Land management and forests are crucial to tackling the concurrent issues of sustainable food production and climate change. Conventional modern agriculture, converting forests and naturally vegetated landscapes to farms and rangelands, contributes significantly to elevate carbon in the atmosphere. Agroforestry systems offer potential for local communities to meet livelihood needs while simultaneously adapting to and mitigating climate change. Data from several studies conducted in nine districts of central Nepal between 2007 and 2017 were analyzed. Forests and agroforestry systems in three central Nepal districts had significantly higher total carbon stocks than agricultural soils (2–5 times) due to high above and below-ground biomass carbon and SOC stocks. The application of improved FYM compost, cattle urine and biochar in four districts increased average SOC by 2.75% over 6 years, translating to an increase of nearly 100 t ha^{-1} in SOC stock. Along with soil quality benefits, biochar and FYM compost improved the yields of soybean, potato, millet and Swertia chirayita yields which were significantly higher than in untreated plots. The flux of N_2O was significantly lower in biochar-amended soil compared to non-biochar. Crop diversification incorporating high-value horticultural and medicinal crops enhance economic returns as indicated by higher benefit-cost ratios for vegetable and Swertia chirayita than for cereals.

Keywords: agroforestry, diversified cropping, climate change, biochar, farmyard manure, sustainable soil management

1. Introduction

Farming communities in the hill and mountain regions of the Himalayas depend to a large extent on the soil and forest resources for their survival and livelihoods. The past several decades have witnessed ever-increasing pressures upon the land-based resources due to increased human and livestock populations. Intensified and commercial agricultural production with increased reliance on chemical fertilizers and pesticides has led to adverse ecological consequences of these unsustainable farming practices. Soil degradation and productivity decline are widespread in many parts of South Asia [1]. Moreover, the impacts of global climate change, causing higher mean annual temperatures and erratic, intense rainfall events, further threaten the livelihoods of small-holder mountain farmers [2, 3]. Therefore, approaches and systems to enhance and sustain production, while also improving resilience to climate change impacts are urgently needed in the Himalayan region. A combination of practices holds considerable potential to provide multiple benefits related to sustainable production along with supporting climate change adaptation and mitigation. These include crop diversification, incorporation of agroforestry species, use of biochar in conjunction with local farmyard manure and the adoption of appropriate soil and water conservation measures [4, 5].

Due to the increasing demands for food and fiber of a growing population, intensification of agriculture is unavoidable. Agricultural intensification can be defined as any cropping or animal husbandry system that increases the intensity or frequency of use of the same parcel of land, for instance, by growing a greater number of crops per year or grazing a higher number of livestock on the same land area than previously [6, 7]. If such intensification of production is done through the adoption of agrochemical use and mono-culture cropping, then over a period of a few decades, the organic matter content and fertility of the soil becomes depleted, leading to diminishing quality and productivity of the land. On the other hand, if intensification can be achieved in a sustainable manner, maintaining ecological balance, then both the livelihoods of farming communities and the environment could be preserved [5, 8, 9].

Agroforestry involves the inclusion of perennial plants on farm land in combination with annual crops. Often trees, such as, fruit or fodder species, are planted within or on the borders of the crop land with other annual cereal or vegetable crops grown in between. Agroforestry systems offer considerable potential as a climate adaptive strategy due to the diverse nature of crops with varying degrees of tolerance to soil, water, nutrient and climatic conditions [5, 10, 11]. Furthermore, the inclusion of perennial crops enables complimentary, as well as, synergistic effects in terms of productivity, while also capturing and storing carbon, thereby contributing to climate change mitigation [12–14]. These systems are likely to be particularly appropriate for mountain farming due to the steep topography, often shallow depth of soils and limited scope for supplemental irrigation.

In addition to diversified cropping strategies through agroforestry, another environmentally friendly and sustainable approach for enhancing production while aiding in carbon sequestration is the use of on-site farmyard manure and locally prepared biochar. While the use of biochar in soils dates back thousands of years in South America and Australia [15, 16], its potential for promoting soil microbial activity, improving plant nutrient availability, increasing water retention and, thereby enhancing production has only gained scientific recognition in recent decades [3, 17, 18]. The beneficial effects of biochar reportedly arise from its high porosity, stability and longevity in soils, serving as a catalyst for increased microbial activity along with high water and nutrient retention capacity [17–19]. In the Himalayan region, biochar prepared from locally available waste biomass feedstock, such as, crop residues and weeds, could offer a low-cost, sustainable option to improve soil quality and livelihoods through increased crop productivity. Layek et al. [9] point out that in the Eastern Himalayas of India, where a type of slash-and-burn agriculture is practiced, the use of biochar could improve soil quality and crop production while also enhancing carbon sequestration. Even greater benefits of biochar to enhance crop yields may be achieved through the application of biochar enriched in nutrients such as nitrogen, phosphorous and potassium [20, 21].

2. Approach and methods

The present chapter is a compilation and re-analysis of the data and salient aspects of several studies carried out in the lower and mid-hills regions of Nepal located in the Central Himalayas. The studies were conducted in nine different districts of the central and mid-western development regions of Nepal during the period from 2007 to 2017. The study districts and locations are shown within the map of the country in **Figure 1**. Brief descriptions of the field trials and studies examined are provided below.

The effects of agroforestry and community-managed forests on the accumulation of carbon in both soil and biomass compared to conventional agriculture were evaluated in a study conducted in three districts of central Nepal. Four field plots each in three land use types, namely, agriculture (AG), agroforestry systems (AF) and community forests (CF), were sampled in Chitwan, Gorkha and Rasuwa districts [22]. These districts represented different agro-climatic zones with Chitwan in the tropical zone (100–300 m elevation range), Gorkha situated in the warm subtropical zone (at 1000–1200 m elevation) and Rasuwa lying in the warm temperate zone (at 1800–2000 m elevation). Soil organic carbon (SOC) and dry bulk density (BD) were determined using dry combustion [23] and core methods [24], respectively, for samples taken at four soil depths down to 1 m or bedrock, whichever was shallower. These included topsoil at 0–0.15 m depth, and three sub-soil layers at 0.15–0.30 m, 0.30–0.60 m and > 0.60 m depths. Using the SOC percent and soil BD, the total soil

Figure 1.
Map of Nepal showing the study districts (shaded in gray). Source: *https://www.worldatlas.com/r/w960-q80/upload/d6/7b/0c/provinces-of-nepal-map.png.*

OC stocks were calculated using Eq. (1) [25]. The above and below-ground biomass carbon stocks were also determined using the allometric method according to Chave et al. [26]. The above-ground tree biomass (AGBT) was calculated using the diameter at breast height (DBH) and the total height of the tree (h) as given in Eq. (2).

The biomass carbon of leaf-litter, herbs and grasses was determined by destructive sampling over an area of 1 m^2, and oven drying in the laboratory; calculations were done according to the formula in Eq. (3). The below-ground root biomass was taken according to the root-to-shoot ratio of 1:5 as suggested in MacDicken [27].

$$\text{SOC stock}\left(\text{t ha}^{-1}\right) = \text{SOC}^{*}\ \text{BD}^{*}\ \text{H}^{*}\ 10^{4} \tag{1}$$

Where, SOC = soil OC %; BD = soil dry bulk density (Mg/m^3); H = thickness of soil layer sampled.

$$\text{AGTB} - \text{C stock}\left(\text{t ha}^{-1}\right) = 0.059^{*}\ \delta^{*}\ \text{DBH}^{2} * \text{h} \tag{2}$$

Where, δ = wood specific gravity (g cm^3); DBH = diameter of tree at breast height (cm); h = tree height (m).

$$\text{LHGB} - \text{C stock}\left(\text{t ha}^{-1}\right) = \ \text{Wf}^{*}\ \text{Wd}^{*}\ 10^{4}/\text{A}^{*}\ \text{Ws} \tag{3}$$

Where, Wf = fresh weight of leaf-litter-herb-grass sample; Ws = wet weight of subsample; Wd = oven dried weight of subsample; A = sampled area (m^2).

The use of sustainable soil management practices, such as, improved compost and farmyard manure (FYM), cattle urine, crop residue mulching and biochar are reported to have beneficial effects on soil quality and fertility. This ultimately translates to increased productivity and higher crop yields, which contribute to greater incomes and better livelihoods for farming communities. A study in four districts of central and mid-western Nepal evaluated the increase in soil organic carbon over a period of 6 years of adoption of these practices by farmers [28]. Soil from 16 farmer fields in Baglung, Kavrepalanchok, Sindhupalchok and Syangja districts was sampled at 0–0.15 m and 0.15–0.3 m depths representing the crop rooting zone. The soils were air-dried, sieved through a 2 mm sieve and analyzed for SOC using the loss-on-ignition dry combustion method. The increase in SOC contents over the six-year period was calculated and the potential for carbon sequestration was determined.

In separate trials in five districts of central Nepal, the application of biochar prepared from locally available crop residues, weeds and grass biomass in combination with FYM compost was studied. The quality of biochar made by pyrolysis at low temperatures (350 to 470°C) using different feed stocks, namely, coffee pulp and husks, leaf-litter/grass, rice straw, Eupatorium sp., and wood sawdust, along with the quality of local FYM compost was determined. The organic C contents, pH, total N, available P, available K and cation exchange capacity (CEC) were determined for biochar, FYM and soil from the farm fields using standard methods. The effects of the local biochar and FYM on soil properties, as well as, crop growth and yields were evaluated for coffee agroforestry and vegetable crops in Bhaktapur, Kavrepalanchok, Lalitpur, Rasuwa and Sindhupalchok districts of central Nepal between 2012 and 2017 [29, 30].

An economic analysis was done to determine the benefits to farm income of cultivating vegetable crops and medicinal plants in smallholder farm fields in Rasuwa district during 2016–2017. Cost-benefit analyses were done for crops including millet (the main cereal crop grown in the district), radish, garlic and Swertia chirayita (medicinal plant). The benefit-cost ratio was calculated to determine the estimated profitability of each crop type.

3. Results and discussion

Based on sampling and quantification of carbon stocks under agricultural fields, agroforestry systems and community forest plots, it was observed that the total carbon stocks accumulated in forests were expectantly highest (by 2 to 5 times), followed by agroforestry systems (approximately 2–3 times higher than agriculture) and least in agricultural lands in three districts in central Nepal (**Table 1**). While soil carbon stocks were variable, with farm fields in Chitwan having significantly higher SOC stocks than the other two districts, biomass carbon stocks were highest for community forests, followed by agroforestry plots. Agricultural plots, on the other hand, did not accumulate biomass carbon as the fields are planted to annual crops which are harvested and all biomass removed each cropping season. The variability of SOC stocks may be due to farming practices, amounts of organic residues returned to the soil and farmyard manure applied, as well as, the soil depth, which was highest in Chitwan district. The influence of organic matter management under different agricultural practices and variability in soil OC contents and stocks have been reported in a number of studies [4, 9, 22, 31, 32].

As shown in **Figure 2**, the distribution of below-ground carbon stocks, including root-biomass carbon and SOC, depended on the depth of sampling. Although the total below-ground carbon stocks were generally higher for forest soils, the difference was not significant when topsoil (0–0.3 m) alone was considered. However, when quantified to greater depths, the below-ground carbon stocks were considerably higher for both community forest and agroforestry plots. This is presumably due to the contribution of root biomass carbon of perennial crops and trees under agroforestry systems and in forests. Increased SOC contents under agroforestry systems have been observed in other studies as well [13, 33–35].

District	Land Use[$]	SOC stock	Biomass C stock	Total C stocks
Chitwan	AG	208.9 ± 37.9	—	208.9
	AF	244.2 ± 44.1	81.3 ± 17.8	325.5
	CF	354.4 ± 129.0	172.3 ± 49.6	526.7
Gorkha	AG	63.0 ± 12.9	—	63.0
	AF	88.0 ± 17.0	63.2 ± 21.7	151.2
	CF	252.5 ± 64.6	75.9 ± 14.9	328.4
Rasuwa	AG	127.1 ± 17.8	—	127.1
	AF	152.5 ± 9.5	262.4 ± 62.8	414.9
	CF	150.0 ± 17.7	420.3 ± 92.8	570.3

[$] *AG = Agriculture; AF = Agroforestry; CF = Community Forest.*

Table 1.
Carbon stocks (t ha^{-1}) under three land uses in three central districts of Nepal.

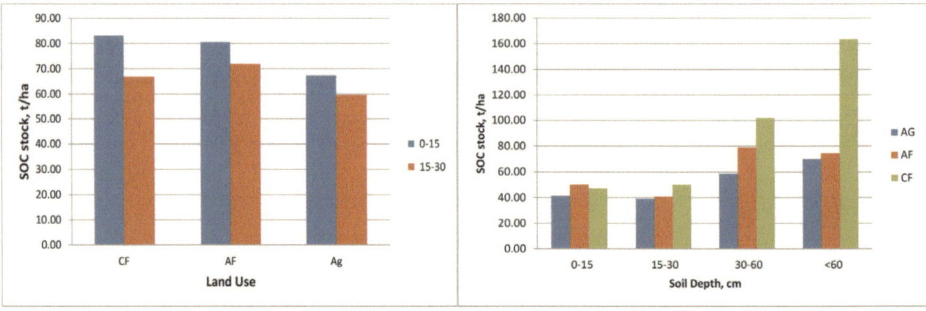

Figure 2.
Distribution of below-ground carbon stocks at different depths within the topsoil (0–0.3 m), in Rasuwa district, left; and the entire soil profile, right, in Chitwan district.

Farm fields in four districts of central and mid-western Nepal, namely, Kavrepalanchok, Sindhupalchok, Baglung and Syangja, were monitored over a period of 6 years to evaluate the effect of sustainable soil management practices that focused on organic and nature-based approaches to increase soil organic matter and fertility. As shown in **Table 2**, the mean baseline SOC contents of the four study districts were low, ranging from less than 1 percent to about 2.3 percent. After 6 years of adoption of the improved FYM/compost and cattle urine application practices on the farm plots, mean SOC contents increased significantly in all districts from about 2.5 to more than 6 percent. This represented a range of about 2-fold to more than 4-fold increase in SOC contents in the farm fields over the 6-year period. Assuming an average soil bulk density of 1.2 Mg m^{-3}, the average increase in SOC stock in the topsoil (0.3 m layer) would amount to nearly 100 t ha^{-1} over 6 years. The results indicate that such sustainable soil management practices can improve not only the quality and fertility of agricultural soils but also contribute significantly to carbon capture and sequestration. Such improvements in soil properties and increased carbon accumulation as a consequence of restorative soil management practices have been shown in numerous other studies [3, 16, 29, 36].

Other studies examining the properties and influence of locally prepared biochar on both the soil and crop production suggested improvements over conventional farming practices. The local biochar was made using a variety of different waste biomass feed stocks such as coffee waste, weeds (Eupatorium sp.), leaf-litter, grass and wood sawdust. Pyrolysis at low temperatures in a locally constructed dual chamber biochar stove over periods ranging from 3 to 10 hours (depending on moisture contents of feed stock) yielded 35 to 50 percent by weight of the resultant biochar as can be seen from **Table 3**. The highest quantities of biochar resulted from leaf-litter and grass, as well as, coffee pulp and rice husk feed stocks at about 50 percent of the initial weight of the biomass.

Biochar has a number of beneficial effects when applied to the soil. This can be seen from the properties of the biochar given in **Table 4**. Biochar tends to be alkaline in nature and hence can moderate the soil pH in acidic soils. The highest pH was observed for biochar made from Eupatorium sp., a local nuisance weed, which resulted in pH greater than 10. The next best feed stock from a pH standpoint was leaf-litter and grass, which yielded biochar with about 9.5 pH. With regard to best overall properties including organic matter, nitrogen and phosphorus levels, Eupatorium along with leaf-litter and grass, as well as, wood sawdust biochar appeared to be favorable (**Table 4**). It should be noted, however, that biochar does

District	Initial SOC %	3-year SOC %	6-year SOC %	Increase in SOC %
Kavreplanchok	0.68	1.36	2.99	2.31
Sindhupalchowk	1.19	1.31	2.45	1.26
Baglung	1.60	3.72	4.96	3.36
Syangja	2.29	2.97	6.37	4.08
Means ± SD	1.44 ± 0.86	2.34 ± 1.20	4.19 ± 1.81	2.75 ± 1.23

Table 2.
Soil organic carbon contents in the crop root zone (0–0.3 m depth) over 6 years of sustainable soil management practices in four districts of Nepal (n = 16).

Feedstock type	Wt, of feedstock (kg)	Wt. of biochar produced (kg)	Fuel used (kg)	Time of pyrolysis (hours)	Biochar yield ratio
Coffee pulp and husk	7.35 kg	3.50	8	10	0.48
Leaf litter and grass	3.4	1.75	1.05	3	0.51
Eupatorium sp.	4.45	1.55	2.5	3.5	0.35
Rice husk	1.41	0.75	4	5	0.53
Wood sawdust	2.5 kg	1	3.5	4.5	0.40

Table 3.
Biochar is produced from various feedstock types under the low-temperature pyrolysis (temperature range maintained at 350–470°C).

not provide nutrients directly to crops, rather, it enhances microbial activity by providing sites with high nutrient and moisture contents for their growth, thereby, enabling the release of nutrients for crop use. Furthermore, due to the high OM contents of biochar and the stability of this OM, biochar can enhance carbon sequestration in soils over extended periods of time. The beneficial effects of biochar on soil properties have also been reported by other researchers [31, 37–40].

Farmers in the hills and mountains of the Himalayas have traditionally used locally prepared farmyard manure or compost as a means of fertilizing and replenishing the soil with organic matter and crop nutrients. In recent decades, however, increasing

Feed Stock Type	pH	Organic Matter (OM) %	Total Nitrogen mg kg^{-1}	Available Phosphorus mg kg^{-1}	Exchangeable Potassium mg kg^{-1}
Coffee pulp and husk	8.41	39.57	4865	1335.5	350
Leaf litter and grass	9.49	35.525	17,465	3848	209.61
Eupatorium sp.	10.25	62.34	4392.5	3259.5	81.87
Rice husk	9.07	25.295	3080	1867	244.25
Wood	8.66	68.54	1120	519	206.2

Table 4.
Biochar properties are produced from five types of biomass feed stocks.

reliance on chemical fertilizers in order to produce commercial crops around the year with an intensified cropping cycle has led to declining soil quality and productivity. Research in several districts of central Nepal, including Lalitpur, Kavrepalanchok, Sindhupalchok, Bhaktapur and Rasuwa, has suggested that the application of a combination of FYM compost and biochar could improve soil quality, while simultaneously enhancing crop yields, particularly those of vegetable crops and medicinal plants.

Analyses of locally prepared FYM compost samples from three districts revealed that this organic form of fertilizer is rich in macro- as well as micro-nutrients (**Table 5**). Apart from having high amounts of organic matter ranging from about 15 to 65 percent, the FYM compost is also high in nitrogen, available phosphorus and available potassium, which are the three most common nutrients required by plants in large amounts. The values in FYM ranged from about 4000 to 7000 mg kg^{-1} for total N, about 5300 to 12,000 mg kg^{-1} for available P and approximately 200 to 300 mg kg^{-1} for available K. These values are considered high to very high as a source of plant available macro-nutrients. Moreover, the analyzed FYM samples also contained considerable amounts of iron, manganese, copper and zinc, which are micro-nutrients required in only trace amounts by crops.

The properties of soils after 2 years of amendment with biochar as compared to soil in control (non-biochar) plots clearly indicate improvements in a number of soil properties (**Table 6**). The SOC content, soil pH and CEC of biochar treated soils were significantly higher than for untreated soils as seen from the analysis of variance F-test and P values. The bulk density was weakly significantly lower ($P < 0.10$) for biochar-amended soil compared to the control. The crop macro nutrients, namely, total N, available P and available K were not statistically significantly different for biochar-amended soil, but the values were, nonetheless, somewhat higher than for the control soil (**Table 6**).

The above results indicate, therefore, that the use of FYM compost prepared on-farm from animal manure mixed with crop residues, leaf-litter and used animal bedding materials serves as an environmentally friendly and sustainable means of restoring soil fertility and overall quality. The enhancement of crop yields due to the application of biochar and organic fertilizers is corroborated by many studies around

Analyzed parameters	Sampled District		
	Lalitpur	Kavrepalanchok	Sindhupalchok
pH	10.77	6.19	8.25
Organic Matter (%)	65.29	15.19	27.23
Total Nitrogen (mg kg^{-1})	7009	4005	6160
Available Phosphorus (mg kg^{-1})	5273	12,061	5333
Available Potassium (mg kg^{-1})	294.1	199.5	196.7
CEC ($cmol_c$ kg^{-1} soil)	67	75	139
Iron (Fe) (mg kg^{-1})	26.60	41.67	2.50
Manganese (Mn) (mg kg^{-1})	17.18	117.9	105.5
Copper (Cu) (mg kg^{-1})	2.64	2.50	3.50
Zinc (Zn) (mg kg^{-1})	10.77	11.50	30.17

Table 5.
Properties of farm yard manure compost prepared locally by farmers.

Soil Property	Control (± SD)	Biochar Tmt. (± SD)	F-test/Signif.
SOC (%)	2.88 ± 0.97	5.07 ± 2.27	25.95**
Bulk density (Mg m^{-3})	1.70 ± 0.23	1.60 ± 0.25	3.47†
Soil pH	5.68 ± 0.43	6.13 ± 0.45	6.09*
CEC ($cmol_c$ kg^{-1})	47.3 ± 11.7	58.6 ± 10.1	6.82*
Total N (mg kg^{-1})	2360 ± 659	2640 ± 426	0.6 ns
Available P (mg kg^{-1})	248 ± 133	302 ± 160	2.16 ns
Available K (mg kg^{-1})	208 ± 39	253 ± 68	1.38 ns

Note: *†, *, ** indicate significance at the 10, 5 and 1 percent levels, respectively.*

Table 6.
Selected properties for biochar-amended and non-biochar (control) soil in three districts of Central Nepal (n = 12).

the world [39, 41–44]. While increasing the rate of biochar application to soil generally led to increased yields, an optimum application rate of 15 t ha^{-1} was proposed by Pandit et al. [45]; as the cost increased with higher rates, the incremental yield did not support the application of higher amounts of biochar.

A field trial on smallholder farm fields in Rasuwa district indicated that the application of FYM compost at a rate of 20 t ha^{-1} mixed with biochar at a rate of 5 t ha^{-1} had significant effects on increasing the yield of a number of crops over the application of FYM alone to the soil (**Table 6**). Yields of mustard seed were highly significantly ($P < 0.01$) different between the two treatments, while those of potato and the medicinal plant Swertia chirayita were significantly different ($P < 0.05$). Although the yields of garlic and radish were only weakly significantly different ($P < 0.10$) between biochar+FYM and FYM alone, the yield values were notably higher for the former by about 25 to 66 percent (**Table 7**).

Another trial in Bhaktapur district also indicated that the combination of FYM compost with low rates of biochar applied to the soil prior to planting led to generally higher yields for crops like garlic, chili, radish and soybean (**Figure 3**). The yield of soybeans grown in soil with biochar+FYM applied was significantly higher, by about 56 percent, than that of soybeans grown in soil with FYM compost only. While the yield differences for the other crops, namely, garlic, chili and radish, were not significant, the yields of the crops grown in soil with biochar+FYM were slightly higher

Crop Type	FYM Compost	Biochar + FYM	P-value/Signif.
	(t ha^{-1})		
Garlic	3.0 2.1	5.0 3.9	0.066†
Potato	6.6 2.4	8.0 3.1	0.012*
Radish	9.0 3.0	11.6 5.2	0.058†
Mustard (seeds)	0.3 0.1	1.3 0.4	0.001**
Swertia chirayita	2.0 1.4	3.04 2.4	0.039*

*†, *, ** indicate significance at the 10, 5 and 1 percent levels, respectively.*

Table 7.
Yields of different crops grown with farmyard manure compost alone and with biochar plus FYM compost in Rasuwa district.

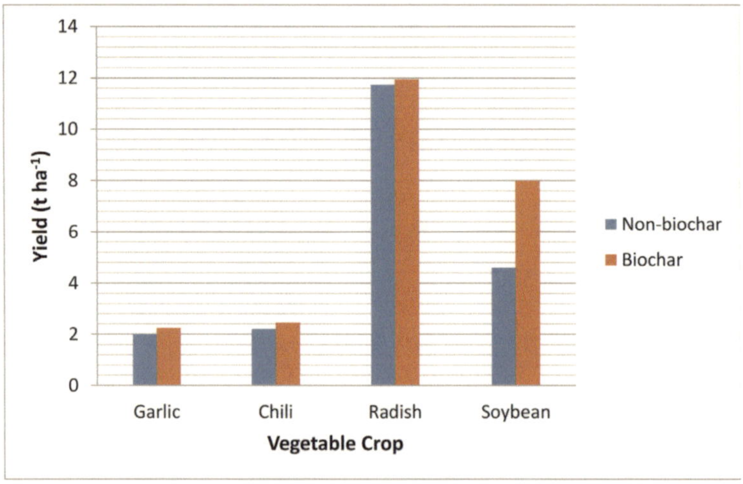

Figure 3.
Yields of vegetable crops grown with and without biochar in Bhaktapur district.

than that for crops grown in soil with only FYM compost (**Figure 3**). Similar results have been reported by researchers in other studies [36, 39, 41, 46]. The application of low doses of biochar enhanced by treating with cattle urine was also found to be effective in increasing yields of a number of vegetable and cereal crops by Schmidt et al. [42, 43].

Apart from the yield benefits of the use of biochar in agricultural and agroforestry soils, another study in Bhaktapur district where biochar was applied to farm plots at low rates (4 to 8 t ha^{-1}) showed a distinct reduction in the emission of greenhouse gases. The monitoring of carbon dioxide, nitrous oxide and methane, using the static chamber method over a period of 2 months during the early growing season, showed lower emissions of the gases from biochar-applied soil than from control plots that received no biochar (**Figure 4**). While the results were not statistically significantly different for CO_2 and CH_4, the emission of N_2O was significantly higher ($P < 0.05$) from soil in non-biochar plots as compared to biochar-amended plots. Other studies have also shown significant reductions in NO_2 and CO_2 from biochar applied soils as reported by Stavi and Lal [4]. On the other hand, increased N_2O emissions from agricultural soils and acidification have been shown to result from intensified crop production with the use of chemical fertilizers [47]. Therefore, the amendment of the intensively cultivated soils with biochar could be a potential measure to alleviate excessive GHG emissions.

Considerable research and scientific evidence have established that agroforestry systems and the incorporation of perennial plant species in agriculture have many environmental and ecological benefits in terms of soil quality, biodiversity, climate mitigation, etc. However, less work has been done evaluating the economic advantages of agroforestry crops, especially, high-value medicinal plants. A study in smallholder farm fields of Rasuwa district in central Nepal compared the total revenues from and costs of cultivating a number of different crops including cereal crops, vegetables and medicinal plants. The results clearly showed that medicinal plant cultivation, such as Swertia chirayita, gave significantly higher returns with a benefit-to-cost ratio of 2.45. On the other hand, growing common cereal crops like maize and millet actually led to a loss, whereas, vegetable crops such as radish were marginally

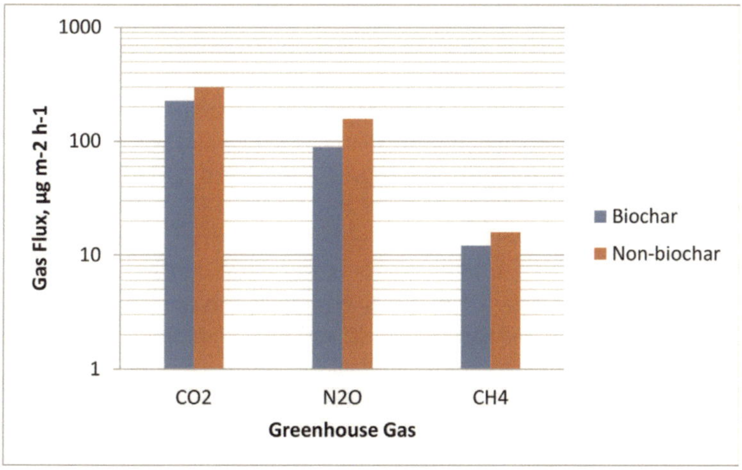

Figure 4.
Emissions of three greenhouse gases from soil with and without biochar amendment in Bhaktapur district.

Crop	Production	Total Revenue	Total Variable Cost	Gross Margin	Benefit:Cost Ratio
	($t\ ha^{-1}$)	(1000 X NPR‡)			
Maize	1.57	62.90	135.73	−72.83	0.46
Millet	0.69	27.52	136.81	−109.29	0.20
Radish	9.83	147.42	136.81	16.81	1.13
Swertia chirayita	393.13	1179.38	481.98	69.40	2.45

‡*Note: Currency exchange rate at the time of study was approx. USD 1 = 102 NPR.*

Table 8.
Cost–benefit analysis for various crops produced by smallholder farmer in Rasuwa district.

profitable (**Table 8**). The economic benefits of diversified cropping and inclusion of horticultural crops have also been reported by Shah et al. [44].

4. Conclusions

The results of a number of field trials conducted in several districts in central Nepal demonstrate that agroforestry, crop diversification and the adoption of various sustainable soil management practices hold considerable potential for enhancing land productivity, increasing crop yields and improving the livelihoods of local farming communities. Agroforestry systems are particularly suitable for mountainous regions and can contribute to climate change adaptation and resilience of hill communities while also increasing SOC accumulation and carbon stocks, thereby, helping to mitigate climate change. Trees outside of forest have been reported to substantially enhance carbon sequestration whether as traditional agroforestry practices or simply trees planted on farms and homesteads [48]. Forests and agroforestry systems typically contained significantly higher total carbon stocks than agricultural soils (by 2 to 5 times) due to the high above and below-ground biomass carbon as well as SOC stocks.

Sustainable soil management practices such as the application of improved FYM compost, cattle urine and biochar in four districts led to an increase in SOC contents by 2.75 percent on average, which could amount to nearly 100 t ha^{-1} of increase in SOC stocks over a 6-year period. In addition to soil quality benefits, the application of biochar and FYM compost improved the yields of a variety of crops with soybean, potato, mustard and Swertia chirayita yields being significantly higher than for untreated plots. Biochar has the potential to contribute to the mitigation of climate change by enhancing carbon sequestration in soil due to its inert nature and longevity in soils, as well as, by helping to reduce emissions of greenhouse gases from soil [49]. The flux of N2O was observed to be significantly lower in biochar-amended soil as compared to non-biochar soil. This was presumably due to the effect of biochar on soil chemistry, moisture conditions and microbial activity [4, 32].

The adoptability of sustainable agricultural practices depends to a large extent on economic returns. Diversifying crops with the inclusion of high-value horticultural and medicinal crops could lead to better economic returns as seen from the relatively higher benefit-cost ratios for vegetable and Swertia chirayita than for cereal crops. Therefore, adequate technical, financial and institutional support to farmers in developing countries and mountainous regions are needed to encourage and promote climate resilient and environmentally sound agricultural and land use practices.

Acknowledgements

The unreserved cooperation and assistance of the hill farmers in our study districts is gratefully acknowledged.

Conflict of interest

The authors declare no conflict of interest.

Author details

Roshan M. Bajracharya[1*], Deepak K. Gautam[2], Ngamindra R. Dahal[3] and Him Lal Shrestha[4]

1 Department of Environmental Science and Engineering, Kathmandu University, Dhulikhel, Nepal

2 Nepal Agroforestry Foundation, Kathmandu, Nepal

3 Nepal Water Conservation Foundation, Kathmandu, Nepal

4 Kathmandu Forestry College, Kathmandu, Nepal

*Address all correspondence to: roshan.bajracharya@gmail.com

References

[1] Lal R. Soil degradation and food security in South Asia. In: Lal R, Sivakumar MVK, Faiz SMA, Rahman AHMM, Islam KR, editors. Climate Change and Food Security in South Asia. Dordrecht, The Netherlands: Springer; 2011. pp. 137-152

[2] Pachauri RK. Climate change and agroforestry. In: Nair PKR, Garrity D, editors. Agroforestry - The Future of Global Land Use. Advances in Agroforestry No. 9. Dordrecht, The Netherlands: Springer Science & Business Media; 2012. pp. 13-16

[3] Bajracharya RM. Regenerative approach for sustainability and climate resilience of mountain agroecosystems. In: Bajracharya RM, Sitaula BK, Raut N, Gurung S, editors. Sustainable Natural Resource Management in the Himalayan Region: Livelihood and Climate Change. New York: Nova Science Publ. Inc.; 2021. pp. 42-56

[4] Stavi I, Lal R. Agroforestry and biochar to offset climate change: A review. Agronomy for Sustainable Development. 2013;**33**:81-96

[5] Bajracharya RM. Sustainable land use, landscape management and governance. In: Shit PK, Adhikary PP, Bhunia GS, Sengupta D, editors. Soil Health and Environmental Sustainability: Application of Geospatial Technology. Switzerland: Springer-Nature; 2022. pp. 423-436

[6] Carswell G. Agricultural intensification and rural sustainable livelihoods: A "think piece". IDS Working Paper No. 64. 2011

[7] Bajracharya RM, Dahal BM. Agricultural Intensification. The Berkshire Encyclopedia of Sustainability: Ecosystem Management and Sustainability. New York, NY, USA: Berkshire Publishing Group; 2012. pp. 7-10

[8] Tilman D, Balze C, Hill J, Befort BL. Global food demand and the sustainable intensification of agriculture. PNAS. 2011;**108**(50):20260-20264

[9] Layek J, Narzari R, Hazarika S, Das A, Rangappa K, Devi S, et al. Prospects of biochar for sustainable agriculture and carbon sequestration: An overview for eastern Himalayas. Sustainability. 2022; **14**(11):6684. DOI: 10.3390/su14116684

[10] Leakey R. Environmental resilience and agroforestry. In: Nair PKR, Garrity D, editors. Agroforestry - The Future of Global Land Use. Advances in Agroforestry No. 9. Dordrecht, The Netherlands: Springer Science & Business Media; 2012. pp. 11-12

[11] Synnot P. Climate change, agriculture, and food security in Nepal - Developing adaptation strategies and cultivating resilience. Report, Mercy Corps Nepal; 2012. 53 p

[12] Bajracharya RM, Atreya K. Carbon sequestration in upland farming systems of the Nepal mid-hills. In: Paper Presented at the National Conference on Environment. Kathmandu, Nepal: Tribhuvan University; 2007

[13] Cardinael R, Chevallier T, Cambou A, Beral C, Barthes BG, Dupraz C, et al. Increase of soil organic carbon stock under agroforestry: A survey of different sites in France. In: 3rd European Agroforestry Conference, Montpellier, France; May 23-25, 2016

[14] Bishaw B, Soolanayakanahally R, Karki U, Hagan E. Agroforestry for

sustainable production and resilient landscapes. Agroforestry Systems. 2022; **96**:447-451. DOI: 10.1007/s10457-022-00737-8

[15] Sandor JA, Eash NS. Ancient agricultural soils in the Andes of southern Peru. Soil Science Society of America Journal. 1995;**59**:170-179

[16] Downie AE, Van Zwieten L, Smernik RJ, Morris S, Munroe RR. Terra Preta Australis: Reassessing the carbon storage capacity of temperate soils. Agriculture, Ecosystems and Environment. 2011;**140**:137-147

[17] Regeneration International. What Is Biochar? San Francisco, CA, USA: Regeneration International; 2018. Available from: http://www.regenerationinternational.org/what-is-biochar/

[18] Novak J, Ro K, Ok YS, Sigua G, Spokas K, Uchimiya S, et al. Biochar's multifunctional role as a novel technology in the agricultural, environmental, and industrial sectors. Chemosphere (Amsterdam, The Netherlands: Elsevier). 2016;**142**:1-3

[19] IBI Biochar. New Jersey, USA: International Biochar Initiative; 2012. Available from: http://www.biochar-international.org/

[20] Sohi SP. Carbon storage with benefits. Science. 2012;**338**:1034-1035. Publ. on-line by AAAS, Washington, D.C.; 2012. Available from: http://www.sciencemag.org/cgi/collection/ecology

[21] Karim AA, Kumar M, Singh E, Kumar A, Kumar S, Ray A, et al. Enrichment of primary macronutrients in biochar for sustainable agriculture: A review. Critical Reviews in Environmental Science and Technology. 2022;**52**(9):1449-1490. DOI: 10.1080/10643389.2020.1859271

[22] Shrestha HL, Bajracharya RM, Sitaula BK. Soil organic carbon and soil properties for REDD implementation in Nepal: Experience from different land use management in three districts of Nepal. In: Bhadouria R, Singh S, Tripathi S, Singh P, editors. Understanding Soils of Mountain Landscapes, Sustainable Use of Soil Ecosystems Services and Management. Amsterdam, The Netherlands: Elsevier; 2023. pp. 267-278. DOI: 10.1016/B978-0-323-95925-4.00002-9

[23] Nelson DWW, Sommers L. Total carbon, organic carbon, and organic matter. Methods of soil analysis. Part 2. In: Chemical and Microbiological Properties, Agronomy Monographs. 2nd ed. Madison, WI, USA: ASA; SSSA; 1982, 1982. pp. 539-579

[24] Blake GR, Hartge KH. Bulk density, Methods of soil analysis part 1. In: Page AL, editor. Physical and Mineralogical Methods, Agronomy Monographs. 2nd ed. Madison, WI, USA: ASA; SSSA; 1986. pp. 425-442

[25] Pearson TRH, Brown SL, Bridsey RA. Measurement guidelines for the sequestration of forest carbon. Gen. Tech. Rept. NRS–18. Newtown Square, PA: U.S. Department of Agriculture, Forest Service, Northern Research Station; 2007. 42 p

[26] Chave J, Andalo C, Brown S, Cairns MA, Chambers JQ, Eamus DE, et al. Tree allometry and improved estimation of carbon stocks and balance in tropical forests. Oecologia. 2005;**145**: 87-99. DOI: 10.1007/s00442-005-0100-x

[27] MacDicken KG. A guide to monitoring carbon storage in forestry and agroforestry projects. In: Forest

Carbon Monitoring Program. Arlington, VA, USA: Winrock International Institute for Agricultural Development; 1997. p. 87

[28] Bajracharya RM, Atreya K. Carbon sequestration in upland farming systems of the Nepal mid-hills. In: Paper Presented at the National Conference on Environment. Kirtipur, Kathmandu, Nepal: Tribhuvan University; 2007

[29] Dahal N, Bajracharya RM. Effects of sustainable soil management practices on distribution of soil organic carbon in upland agricultural soils of mid-hills of Nepal. Nepal. Journal of Science and Technology. 2012;**13**(1):133-141

[30] Gautam DK, Bajracharya RM, Sitaula BK. Effects of Biochar and farm yard manure on soil properties and crop growth in an agroforestry system in the Himalaya. Sustainable Agriculture Research. 2017;7(4):74-82. Available from: https://www.ccsenet.org/journal/index.php/sar/article/view/69546

[31] Khan N, Shah Z, Ahmed N, Ahmad S, Mehmood N, Junaid M. Effect of integrated use of biochar, FYM and nitrogen fertilizer on soil organic fertility. Pure and applied Biology. 2013; **2**(2):42-47

[32] Semida W, Beheiry H, Setamou M, Simpson C, Ali T, Rady M, et al. Biochar implications for sustainable agriculture and environment: A review. South African Journal of Botany. 2019; **127**:333-347. DOI: 10.1016/j.sajb.2019.11.015

[33] Dahal N, Bajracharya RM, Wagle LM. Biochar effects on carbon stocks in the coffee agroforestry systems of the Himalayas. Sustainable Agricultural Research. 2018;7(4): 103-114. DOI: sar.v7n4p103

[34] Ofosu E, Bazrgar A, Coleman B, et al. Soil organic carbon enhancement in diverse temperate riparian buffer systems in comparison with adjacent agricultural soils. Agroforestry Systems. 2022;**96**:623-636. DOI: 10.1007/s10457-021-00691-x

[35] Noponen MRA, Healey JR, Soto G, Haggar JP. Sink or source-the potential of coffee agroforestry systems to sequester atmospheric CO_2 into soil organic carbon. Agriculture, Ecosystems and Environment. 2013;**175**:60-68. DOI: 10.1016/j.agee.2013.04.012

[36] Fan M, Shen J, Yuan L, et al. Improving crop productivity and resource use efficiency to ensure food security and environmental quality in China. Journal of Experimental Botany. 2013;**63**(1):13-24

[37] Pudasaini K, Ashwath N, Walsh K, Bhattarai T. Biochar improves plant growth and reduces nutrient leaching in red clay loam and sandy loam. In: Proc. Int'l. Conf. Water, Food Security and Climate Change in Nepal. Spl. Issue, J. Water Energy and Environ. Kathmandu, Nepal: HydroNepal; 2012. pp. 86-90

[38] Ścisłowska M, Włodarczyk R, Kobyłecki R, Bis Z. Biochar to improve the quality and productivity of soils. Journal of Ecological Engineering. 2015; **16**(3):31-35. DOI: 10.12911/22998993/2802

[39] Njoku C, Uguru BN, Chibuike CC. Use of biochar to improve selected soil chemical properties, carbon storage and maize yield in an ultisol in AbakalikiEbonyi state, Nigeria. International Journal of Environmental and Agricultural Research. 2016;**2**(1): 15-22

[40] Pandit NR, Mulder J, Hale SE, Martinsen V, Schmidt HP,

Cornelissen G. Biochar improves maize growth by alleviation of nutrient stress in a moderately acidic low- input Nepalese soil. Science of the Total Environment. 2016;**625**:1380-1389

[41] Akhtar SS, Li G, Andersen MN, Liu F. Biochar enhances yield and quality of tomato under reduced irrigation. Agricultural Water Management. 2014; **138**:37-44

[42] Schmidt HP, Pandit BH, Cornelissen G, Kammann CI. Biochar based fertilization with liquid nutrient enrichment: 21 field trials covering 13 crop species in Nepal. Land Degradation and Development. 2017;**28**(8): 2324-2342. DOI: 10.1002/ldr.2761

[43] Schmidt HP, Pandit BH, Martinsen V, Cornelissen G, Conte P, Kammann CI. Four-fold increase in pumpkin yield in response to low dosage root zone application of urine-enhanced biochar to a fertile tropical soil. Agriculture. 2015;**5**(3):723-741. DOI: 10.3390/agriculture5030723

[44] Shah AK, Kori AK, Kumar K, et al. Yield performance and economic evaluation of mustard varieties under mango based Agri-horticulture practice in semi-arid tropics. Agroforestry Systems. 2022;**96**:651-657. DOI: 10.1007/s10457-021-00712-9

[45] Pandit NR, Mulder J, Hale SR, Zimmerman AR, Pandit BH, Cornelissen G. Multi-year double cropping biochar field trials in Nepal: Finding the optimal biochar dose through agronomic trials and cost-benefit analysis. Science of the Total Environment. 2018;**637-638**:1333-1341. DOI: 10.1016/j.scitotenv.2018.05.107

[46] Schulz H, Glaser B. Effects of biochar compared to organic and inorganic fertilizers on soil quality and plant growth in a greenhouse experiment. Journal of Plant Nutrition and Soil Science. 2012;**175**:410-422

[47] Raut N, Dorsch P, Sitaula BK, Bakken LR. Soil acidification by intensified crop production in South Asia results in higher N2O/(N2 + N2O) product ratios of denitrification. Soil Biology and Biochemistry. 2012;**55**: 104-112

[48] Skole DL, Mbow C, Mugabowindekwe M, Brandt MS, Samek JH. Trees outside of forests as natural climate solutions. Nature Climate Change. 2021;**2021**(11):1013-1016. DOI: 10.1038/s41558-021-01230-3

[49] Gogoi N, Sarma B, Mondal SC, Kataki R, Garg A. Use of Biochar in sustainable agriculture. In: Farooq M, Pisante M, editors. Innovations in Sustainable Agriculture. Cham: Springer; 2019. DOI: 10.1007/978-3-030-23169-9_16

Chapter 2

Fire and Climate Change in the North American Great Lakes Pine Transition Forest

Steven I. Apfelbaum and Alan Haney

Abstract

This chapter explores potential changes in fire regimes and biodiversity given projected changes in climate in the Great Lakes Pine Transition Forest (GLPF), a portion of the southern boreal forest. We have studied how forest structure, composition, and biodiversity changed over nearly 500 years by comparing communities on comparable edaphic locations of various ages since stand-replacing disturbance. Our interpretation of probable future changes is based on how these ecosystems have reassembled after wildfire, blowdown, and insect disturbances in the past and projections of climate change in the next 75 years. Rapid climate change is now introducing different conditions from those to which this ecosystem was historically adapted, and therefore, projections of future change are more subjective.

Keywords: boreal forest, great lake pine forest formation, forest carbon, peat carbon, wildfire climate risk, fire and climate change

1. Introduction

The boreal ecosystem is one of the most extensive with a circumboreal distribution accross North America, Europe, and Asia. This is also one of the largest fire maintained forested systems on earth. In this chapter we evaluate future trajectories for the Great Lakes Pine Transition Forest (GLPF), a southern extension of the Boreal forest system bordering the Great Lakes in North America. This forested system has an long fire record from tree ring, soil carbon-charcoal, and palenological analyses by others. In this chapter we used our nearly forty years of field measurements of forest structure, standing crop biomass, among other measurements to make informed projections of future changes in this ecosystem [1, 2].

2. Ecosystem description and historic response to disturbance

The GLPF, as with most of North America's ecosystems, was historically adapted to periodic fire. With Indo-European settlement in the mid-to-late 1700s essentially all ecosystems of North America were altered increasingly as populations and technology advanced. Those not directly affected by timber harvests or forest removal, cultivation, or hydrologic alterations were altered by fire suppression and introduction of exotic species. Rapid climate change threatens to further change ecosystems, often unstable because of previous stresses, including loss of species. Although the GLPF has not lost many species, if any, the relative abundance and relationships have been altered. Ecosystems such as the GLPF, adapted to fire regimes that changed slowly over nearly 10,000 years, now must adapt to changes in fire regimes in a matter of several decades. Moreover, increased fire frequency and intensity, that now appears inevitable, will contribute more greenhouse gases emissions in a negative feedback loop.

The GLPF lies between the Boreal Forest Region in North America, which extends from Newfoundland to Alaska to the Great Lakes–St. Lawrence Forest Region, as defined by Rowe [3]. Circumpolar boreal ecosystems, of which the GLPF is a segment, are the largest remaining intact forested ecosystem on the planet [4]. The GLPF is distinguished most obviously by presence of red pine (*Pinus resinosa*) that sometimes occurs in nearly pure natural stands, but more often, when present, is a co-dominant. Eastern white pine (*Pinus strobus*) is also a signature species of the GLPF, although its range extends much farther south, well beyond the boreal forest biome. Otherwise, most of the trees common to at least the central boreal forest of North America coexists in the GLPF. Typical associates of red-and-white pine are jack pine (*Pinus banksiana*), balsam fir (*Abies balsamifera*), white-and-black spruce (*Picea alba and P. mariana*), paper birch (*Betula papyrifera*), tamarack (*Larix laricina*), northern white cedar (*Thuja occidentalis*), quaking aspen (*Populus tremuloides*), and balsam poplar (*Populus balsamifera*), proportions varying according to edaphic characteristics, and type and time since the disturbance that gave rise to the current community.

A classification of 13 upland plant communities found in the GLPF was developed by Ohmann and Ream [5] and Grigal and Ohmann [6]. Classification of lowland communities was developed by Heinselman [7] and Dean [8]. These eight lowland community types plus the 13 upland community types provide a baseline for predicting changes associated with climate change. Both the upland and lowland communities have abundant carbon present in standing vegetation and peat substrates, much of which might be released by increased fire.

The tree species listed above are dominant or codominant in all except three community types: lichen communities on exposed rocky outcrops; alder-willow shrub communities, primarily along streams or in frequently flooded areas; and marsh and open muskeg. A fourth wetland community is dominated by black ash (*Fraxinus nigra*) and American elm (*Ulmus americana*) that is becoming less common because of disease. The composition, structure, and diversity of all upland communities are strongly influenced by the intensity and extent of wildfire [1].

Lowland communities, historically, burned, usually after exceptionally dry periods, often several successive years with below-average precipitation. Under conditions of low humidity and high wind, crown fires can sweep across lowland communities even without prolonged drought. With drought, even substrates burn. In extreme cases, peat-filled depressions can become open ponds. Despite only covering around 3% of land surface area, peatlands are responsible for storing up to one-thirds of the world's soil carbon [9, 10]. This is twice as much carbon as in all the world's forest biomass combined. Peatlands play a critical role in filtering and storing water. They improve water quality and reduce floods [11, 12]. They are home to many rare and unique species. Following fire, pines, especially jack pine, typically dominate upland sites and spruce and fir or aspen dominates sites with deeper soil and more reliable moisture.

Strong straight-line winds are a secondary disturbance in the GLPF. High wind often results in extensive blowdown. Aspen is a vigorous root-sprouter, and when present in predisturbance communities, will dominate after blowdown or clear-cutting, especially on deeper soils with adequate moisture [1, 13–15]. Aspen, with wind-blown seed also can invade extensive burns from surviving parent trees that may be located well outside the burn. Aspen is intolerant and relatively short-lived; consequently, following blowdown, aspen dominance slowly gives way to more tolerant spruce and fir, although surviving aspen may persist as a scattered overstory for 100 years or more. In the long-term absence of disturbance, spruce-fir forests eventually replace aspen on deeper and more mesic sites. Where seed sources are present, white and red pine may also become established and persist into old-growth, 250 or more years [1]. Red pine more often regenerates after fire, whereas the more tolerant white pine may regenerate after fire or wind-throw. Both white and red pine develop thick bark that can withstand fire after 30 to 50 years, and they can survive for 400 or more years under favorable conditions, creating an overstory above spruce, fir, or jack (pine). Such old-growth communities are rare, in part not only because the commercial value of mature white-and-red pine led to widespread harvesting but also because intense fire associated with periodic drought regenerated forest communities, with or without white-and-red pine depending on whether parent trees survived long enough afterward to release viable seed.

Our studies have focused on vegetation and breeding birds [16] and blowdown [13–15], and corresponding breeding bird communities and species associated with the vegetation structure and compositional responses [17–20]. A synopsis of some likely changes include a decline in diversity of bryophytes and other understory vegetation is likely. Breeding bird diversity in jack pine-black spruce communities declines as these even-aged communities mature. As for several postfire years, diversity is higher, frequent fire might favor bird diversity. After intense stand-replacing crown fires in upland jack pine communities, a homogeneous jack pine or jack pine-black spruce is regenerated. The relatively low diversity of the bird community, dominated primarily by bark-gleaning and tree-foliage-searching foraging guilds, in the rather homogeneous jack pine forest stand is replaced by species that include ground brush foragers that feed on seeds and insects, tree foliage and brush foliage searchers that feed on insects, fly catching birds that sally forage for flying insects across the openings and open for structure of a newly burned site, woodpeckers that nest in the dead and dying trees and forage forest bark beetles, and numerous raptors that arrive in search of grouse another bird foraging on the lush insect and seed production at the ground story.

The presence of abundant aspen saplings provides food and cover for much wildlife, especially important for charismatic species such as moose and beaver. Food chains in young aspen is indicated by heavily browsed saplings and presence of fecal material of moose and snowshoe hare, associated with fecal material of wolves which contain the hair from these herbivores. Northern Goshawk (*Accipiter gentilis*) localize their nest around recent burns, where they actively have been documented to pursue their primary foodstuff, ruffed grouse (*Bonasa umbellus*) and where we have observed and documented their nesting [18, 21, 22].

3. Past and future regional climate

With projected climate changes, both blowdown and fire will become more frequent, and likely will become more extensive given higher temperatures and more extreme climatic events, especially thunderstorms. Although there are excellent records of climate change over the past 50 or more years, and very good models of how climate will change in the next 50 years, the relationship between climate and fire characteristics is much more nebulous. Much of the anticipated change in climate and projections of associated impacts for the Great Lakes Region is provided in a report (An Assessment of the Impacts of Climate Change on the Great Lakes) by the Environmental Law and Policy Center [23] prepared by 18 climate scientists with expertise in the region. That report, however, did not include a discussion of how climate change might affect forest ecology and fire.

Climate is changing faster in the Great Lakes region than averages over the rest of United States. Averaged over 1985–2016 relative to 1901–1960, in the Great Lakes Basin the temperature has increased 1.6°F (0.53°C). This is faster than the 1.2°F (0.4°C) increase for United States [24]. Teasing these trends apart reveals an average increase across Wisconsin, for example, of over 4°F (1.3°C) in winter and 1°F (3.0°C) in summer. Total annual days of ice-cover for Lake Mendota in Madison, Wisconsin, where good records have been kept since 1860, shows a steady decrease from around 120 days to a present five-year running average of around 80 days. Arguably more important than averages is the increase in year-to-year variation. For example, the maximum percent of Great Lakes covered by ice has remained very close to the 50-year average of about 50%. The year-to-year variation, however, has increased substantially, with higher and lower extremes. Averaged over the same time periods, annual precipitation increased in the Great Lakes Basin by 10%, but much of the increase fell in winter and early spring where it has less effect on fire regimes. Summer precipitation in the Basin is predicted to decrease by 5–15% by the end of the century [25]. Again, the variation has increased with greater and lower amounts and more prolonged droughts.

Globally, extreme weather events are expected to become more common. In North America, heat waves and drought have become more frequent since the 1960s [24]. When comparing average annual daily mean temperatures for the 1976–2005 period to that predicted for the 2070–2099 period, an increase of 5.8–10.1°F (1.9–3.4°C). is projected for the Great Lakes Basin [23]. The range reflects two different scenarios; the lower scenario is based on optimistic reduction in greenhouse gas (ghg) emissions while the higher scenario is based on a projection of current trends in ghg emissions. Extremely warm days (those with temperatures exceeding 90°F (32.2°C)) are projected to increase by 17–40 days by the end of the century, the range reflecting the low- and high-ghg emission scenarios. Wintertime warming is expected to increase proportionately more than summertime high increases, as documented for Wisconsin above. The frost-free period will increase substantially. Days with minimal temperature exceeding 32°F (0°C) in the Great Lakes Basin are projected to increase by 27 and 42, with the low and high scenario models, respectively. These highly probable climate changes can only result in more frequent and more severe fire.

Based on historic fire patterns, conservative fire projections can be offered. Historic fire return intervals vary depending on cycles of drought and landform position, but average 70–110 years [1]. It has been rare that any upland forest escaped fire for more than 250 years. A few pockets of old growth up to 450 years have survived when protected by sheltering topography (e.g., draws, lakes, rock escarpments). There can be little doubt that with even the most optimistic climate projections for higher summer temperature, drier conditions, and more severe storms for the next 50 years, fire frequency will increase. Because this region is often where cold dry arctic air masses clash with warm, moist Gulf air masses, thunderstorms with substantial lightning are typical and, not infrequently, derechos develop with hurricane-force straight-line winds. Even under current conditions, winds associated with thunderstorms commonly exceed 60 mph (96.6 km/hr).

Lightning in the region is primarily generated by frontal and convective activity [26]. On average 27 thunderstorms occur annually in the Great Lakes region. Lightning occurs mostly mid-May to mid-September. Although forest conditions become increasingly dry as the growing season progresses, exposure to lightning decreases.

Prolonged dry periods, often extending over several consecutive years, have led to more widespread and more severe fires in the past [1]. Such drought periods are likely to continue and become more severe. Droughts may become so severe as to be the cause of the death of trees and dramatically reduce lake levels. Even one or 2 months with little to no precipitation during spring or early summer can set the stage for severe fires [1]. For example, the region experienced below normal precipitation from 1929 through 1936 and extreme high temperatures were recorded in 1936 leading to outbreak of many severe fires in the region. It is during such droughts that lowland communities often burned.

4. Projections of how climate change will alter the GLPF

Heinselman [1] developed forest stand-origin maps of extensive portions of the GLPF using dendrochronology to determine forest tree ages. Fire scars and tree growth ring measurements have been used to identify periods of drought associated with wildfire patterns. Swain [27] used charcoal and other pyrolyzed material in annual lake varves to examine fire frequencies. These data were used to document return intervals and relationship of different communities to fire. Fire with return intervals of 65–100 years on upland sites have led to jack pine and aspen dominance over much of the region. In fire-protected draws such as lake margins and wetlands, white and red pine, spruce, and many species of shrubs may escape fire for 100–200 or more years. Many wetlands burned only during the most severe droughts [1].

Heinselman [1] found many, but not all, fire scars to be associated with major fire years, where intense crown-fires resulted in stand-replacement. However, during intervening years, ground fires that under-burned forests of multiple age-classes and origins, confused the interpretation of stand ages. For example, during droughts, peat becomes combustible and, once ignited, can retain fire for months, even into winter. The resulting patterns of vegetation largely reflect the combustibility of fuel, especially the difference between those fuels able to carry crown-fires and wetlands or protected areas more prone to lighter, ground fires.

5. Fires and biodiversity

Fire largely shapes vegetation composition which, in turn, largely controls the breeding and overwintering birds, mammals, insects, amphibians, and reptiles that persist in each community. In addition to biodiversity, fire and postfire succession shapes the structure of the vegetation [1]. Wind or insect outbreaks also influence which dominant species of trees and shrubs will develop, but fire can, and often does, trump those disturbances.

Addressing the most common community types, we can project the responses likely with climate change. Based primarily on Heinselman' s work [1, 28, 29], the most common community types in the GLPF are fir-birch (16%), jack pine-black spruce (11%), black spruce bog (7%), maple-aspen-birch-fir (6%), aspen-birch (6%), jack pine-fir (6%), and black spruce-feather moss (6%). Here, we summarize forest community changes likely with the projected climate changes.

1. *The fir-birch communities*. This community is commonly affected by periodic outbreaks of spruce budworm that cause die-off of balsam. The community is most often found on lower slopes near lakeshores, streams, and wetlands. While this community type occupies topographic positions somewhat protected from fire, it is vulnerable. Most have developed over 100 years since fire. Higher-fire frequency and greater intensity will tend to greatly reduce the fir component and lead to a birch-dominated younger forest type.

2. *Jack pine-black spruce communities*. This common community type occurs mostly on ridges and slopes. Most are even-aged stands that originated after dry periods and widespread fire in 1863–64, 1875, 1894, 1903, 1910, and 1918. Biodiversity in this community type is lower than in most others. This may be a community, however, that is best adapted to increasing frequency and severity of fire and is the community type most likely to experience more fire. Both jack pine and black spruce reseed after fire. Jack pinecones are highly serotinous, but black spruce also has persistent cones that frequently contain viable seeds that are released after being exposed to crown fires. Jack pine trees as young as half a dozen years will produce cones. In the unlikely long-term absence of fire, jack pine will gradually die and be replaced by the more tolerant black spruce. Even-aged jack pine-black spruce stands will continue, but average age likely will be reduced to 25–50 years.

3. *Black spruce bogs and black spruce-feather moss community types*. These communities are either an acidic bog type or better drained with feather moss ground cover. The bog type usually has a bryophyte-dominated ground cover but may have an understory of speckled alder (*Alnus rugosa*) and associated species with greater diversity. Tree growth is slow, but this type is somewhat protected from fire by its topographic position and wet substrates. Peat accumulation under this and other lowland forest communities varies in depth. In depressional locations, peat can be 5–20 m in depth, with shallower deposits (i.e., <1 m in depth) in upland communities, and along margins of depressional landforms [9, 11]. During prolonged dry periods, wildfires often eliminate most of the shallower peats and burn to varying depths in the deeper depressional deposits. During normal or wetter conditions, crown fires readily burn over the peatland forest cover without combusting the peat. With changing climate, fire is likely to result in younger black spruce-dominated forests with less peat. Deeper peat will also be reduced, but likely will persist in lowlands that hold water through drought periods.

This community type often includes extensive shallow peat substrates with a shallow 15 cm depth which contains ~60–90 tons of carbon/hectare [30]. Using these tonnages for estimation and the acreage of GLPF and southern boreal forest in Ontario (26 million ha of forested peatlands), this conservatively suggests ~1500–2300 million tons of carbon to be present in the peat. If this burns, it releases 1.65 to 1.8 times this, or 2400 to 4140 million tons of CO_2 equivalent, or 0.2 to 0.4 gigatons of GHG emissions. More emissions would be expected if fires burned deeper peat deposits with increased summer heating and drought. Peatlands in this small example from Ontario represent a very small percentage of the estimated 600,000 million metric tons of peat estimated to be present across on earth [9, 11].

4. *Maple-aspen-birch-fir and aspen-birch community types*. These communities occupy more mesic sites than jack-pine cover types and can occur in any topographic location but are most common on lower slopes or more level areas. In addition to dominant aspen and birch, they have an understory of co-dominant fir. Many originated after old growth white pine were harvested. In areas not logged, most originated from fires in 1864, 1875, 1894, and 1910, or if after logging, from slash fires between 1895 and 1930. The mesic nature of these communities and the dominance of deciduous trees give these forests some protection from climate change. With increasing drought and high fire intensity, however, we can anticipate increased dominance of birch and growing presence of jack pine.

5. *Jack pine-fir communities*. This community type originated following fire, most often, between 1854 and 1925. The community is dominated by even-aged, pole-sized jack pine and scattered aspen with various understory and codominant fir with some black spruce. These are usually species-rich communities on mesic-to-dry sites. With climate change, jack pine is strongly favored to increase dominance, with less fir and more black spruce. Understory vegetation and breeding bird diversity will decrease.

6. Future fire and climate relationships

At the scale of the landscape, the following fire and ecosystem-scale changes are likely:

1. *More winter/spring precipitation*. The GLPF region absorbs solar radiation that quickly warms the land. Increased winter and spring precipitation may have little effect on fire regimes of the GLPF, but especially with warmer and drier summer conditions, may further cause a shift of fire to later in the season. Increased runoff with the overall increase in precipitation, coupled to fire disturbance, may increase the saturated zone around wetlands, and increase the duration of flooding, thereby reducing the combustibility of lowland forests. This may also increase their effectiveness as fire breaks, reducing the size and rate of spread of spring and early summer wildfires, thereby reducing burn patch sizes and continuity across the landscape.

2. *Ignition of wildfire*. Intensity of spring storms when the contrast between colder Canadian air masses and warmer Gulf air masses are greatest will increase the

incidence of lightning ignition. Although thunderstorms are less frequent in summer, they do occur and with warmer, drier conditions, lightning ignition will likely increase.

3. *Drier, warmer summers, and more extreme droughts*. Larger and more intense wildfires are quite likely, along with shorter fire return intervals. This will result in younger more even-aged forests, especially on better-drained topography. Stump and root-sprouting trees and shrubs (e.g., aspen, birch, mountain maple, and alder) will be favored, especially on mesic sites where jack pine is less aggressive. Peat and organic material of any nature will likely be decreased. With more intense storms and less-organic protection, uplands will be subjected to increased erosion, with resulting siltation and nutrient enrichment of lakes and streams.

4. *Old-growth remnants, already rare, will become even less common.* White pine, especially, will be disfavored as jack pine increases. Older (>300 years) and mid-age stands (150–300 years) will be replaced by younger stands that will be more homogeneous. As old growth stands and trees decline, so too will the nesting opportunities for Bald eagle, Osprey, barred owl, Great gray owl, ravens, chimney swifts, and such species as purple finch and red crossbills. Pine martens, fishers, bobcat, black bear, among other wildlife will also decline. Woodland caribou, moose, and timber wolves may be favored in the increased graminoid, and lichen-covered open ground.

7. System changes

Durability and longevity of carbon on the landscape in standing or downed woody biomass will decrease. GHG's emissions, in time, will decline as future fires burn through younger stands with less woody biomass, and most vulnerable peat in wetlands is burned.

Biodiversity, especially of understory species and neotropical birds will continue to go down. Habitat loss and climate change are already reducing migrant bird species. A recent report by the National Audubon Society estimates that two-thirds of North American birds are endangered species of extinction by the end of the century. In the GLPF, birds that nest or feed in open ground and that utilize brush and young postfire jack pine stands will be less impacted by changing fire regimes. Mourning Warblers, Chestnut-sided Warblers, Yellow-rumped warblers, and Ruffed Grouse are examples of birds that will be favored [18, 19]. Species that will decline during the replacement of mid-age and older stands by these younger stands will be the tree-foliage-searching guild, such as Solitary and Red-eyed Vireo, Parula, Canada and Black-throated Green Warblers, Ruby-crowned Kinglets, Spruce Grouse, Hooded Merganser, and Northern Goshawk [21, 22]. Woodpeckers, including northern black-backed and arctic three-toed woodpeckers, pileated and downy and hairy woodpeckers are all envisioned to increase and colonize burned settings with dead standing trees. Flycatchers are expected to increase as the dense stands open after fire with canopy mortality, foraging species that sally (e.g., olive sided flycatcher, least flycatcher), while species that forage on-wing (e.g., night-hawks, chimney swifts, tree swallows) would increase.

Plant species shifts will occur, but in our experience, no species is lost from the landscape following fire [2, 16]. Native graminoids (*Carex, Calamagrostis*, etc.) will increase as will species of more open sites such as berries (blueberry-*Vaccinium*,

Blackberry and raspberry (*Rubus* spp), and bearberry (*Arctostaphylos* sp). Mountain ash (*Sorbus americana*), mountain maple (*Acer spicatum*), Haws (*Viburnum* spp) are also likely to increase. Species of bryophytes, especially mosses, leafy liverworts, and ferns of deep-forested settings, especially arboreal species found on the trunks of living trees are likely to decline.

8. Climate and fire

Climate change can acerbate effects that have not previously been considered. One such feedback loop is the changing fire regime and ghg emissions from carbon contained in the standing and downed timber and peat substates. To understand the differences in the magnitude of the GHG emissions as an example, we compare the carbon present in the forest cover and forested peatlands in Ontario [11].

Peatland carbon. Boreal forests comprise 83% of all North American forests. The above Ontario example (covering 26 million ha) only represents ~30% of all peatlands in Ontario's boreal region [9, 11]. Just in this region, as fires become more frequent and intense, 24–41 million tons of CO_2 equivalent could be released to the atmosphere from combusted peats.

Forest Carbon. Ontario forests containing 193–214 mega-tons of carbon/ha (MgC/ha) accrue additional carbon at rates of 2.1–3.7 MgC/ha-yr [30]. An additional 40% (or 193 MgC/ha × 1.40) is found in the roots, as below-ground carbon (or 77 MgC/ha). Using the same Ontario forested peatlands acreage as above, >7 million tons (5,043,090,000 MgC above-ground, and 2,012,010,000 MgC below-ground in roots) could be combusted with increased frequency and intensity of wildfires in this Ontario landscape.

9. Summary and considerations

Changes in climate and wildfire regimes are predicted to result in large impacts in the GLPF and release of carbon may have even more profound impact, especially when scaled to the entire boreal ecosystem. Significant increases in fire intensity, frequency, and the acreage burned are predicted in this ecosystem with climate change predictions [11, 12, 31].

Release of carbon from the biome may well influence future climate change, altering climate significantly more than the current predictions. This suggests several considerations for forest/ecosystem management as follows:

1. *Preemptive prescribed burning of expansive upland jack pine communities.* In the larger GLPF/boreal ecosystem, jack pine and spruce communities are the most widely distributed and are some of the most fire vulnerable of all communities. Where cool "under-prescribed burning" can be conducted to reduced ladder fuel loads and combustible fuel levels, this could reduce the future of crown fires that can spread rapidly over the remaining uplands and into forested lowlands.

2. *Preemptive prescribed burning to protect age class frequency distributions for mid and old growth forest systems*. Already rare, mid, and older growth forest systems will be increasingly stressed and will decline under the climate change predictions. Biodiversity of these older systems will shift to younger growth forest systems

and with this a decline in regional diversity is likely. The use of preemptive prescribed burning in forest communities around the locations with mid- and older-age communities may help reduce the risk of spread of regional fires to these older communities.

3. *Preemptive prescribed burning to protect age class frequency distributions for mid- and old-growth forest systems.* Already rare, mid, and older growth forest systems will be increasingly stressed and will decline under the climate change predictions. Biodiversity of these older systems will shift to younger growth forest systems and with this, a decline in regional diversity is likely. The use of preemptive prescribed burning in forest communities around the locations with mid- and older-age communities may help reduce the risk of spread of regional fires to these older communities.

4. *Preemptive mitigation of dewatering of peats and landscapes from mining, development, and water and forest management projects.* Dewatering of landscape surface and ground water resources is a common intentional and inadvertent activity associated with mining and development. Many silvicultural activities involve intentional dewatering to gain growing season access to timber resources. Restoration of hydrology to reverse the effects of dewatering should be considered to reduce the increases in the normal year drying (and drought year) effects and increases the risk of wildfire and its spread and intensity.

5. *Rare species and community-specific plans for mitigation of fire frequency and intensity changes will be needed.* Some species and communities, particularly fire microsites, springs, and seeps, and in fens and other rare peatland resources will decline with predicted climate change.

6. *Monitoring to understand trajectory of effects and changes.* As one of the largest global scale forest and peat-rich areas on earth, a real-time monitoring program should be put into place to measure and monitor drying trends, fire-risk trends and locations with the aim of emulating natural fire regimes and ensure that the above outcomes, among others, are built into management plans.

7. *Creation of assessments and action plans to address each these considerations*. (a) Reducing the potential for the spread of wildfires with appropriate economic incentives to create landowner participation. (b) Protection of loss of mid- and older-growth communities. (c) Restoration of landscape hydrology and developing alternatives for water sourcing. (d) Rare species and microsite mapping and management of biodiversity refugia to reduce losses and extirpation and (e) Establishment of on-the-ground monitoring and remotely sensed imaging to target mitigation and remedial restoration strategies for restoration and management of the GLPF/boreal ecosystem.

Author details

Steven I. Apfelbaum[1*] and Alan Haney[2]

1 Applied Ecological Institute, Inc., Brodhead, WI, USA

2 University of Wisconsin, Stevens Point, WI, USA

*Address all correspondence to: steve@aeinstitute.org

References

[1] Heinselman M. The Boundary Waters Wilderness Ecosystem. Minneapolis: University of MN Press; 1996. p. 334

[2] Apfelbaum SI, Haney A, Wang F, Carlson J. Old-growth southern boreal forest stability response to a stand-replacing wildfire. Natural Areas Journal. 2017;**37**(4):474-488

[3] Rowe JS. Forest Regions of Canada. Canadian Dept. of Environment: Canadian Forest Service Publ; 1972. p. 1300

[4] Collinson AS. Introduction to World Vegetation. Springer; 1978

[5] Ohmann LF, Ream RR. Wilderness Ecology: Virgin Plant Communities of the Boundary Waters Canoe Area. U.S. Forest Service res. Paper NC-63; 1971

[6] Grigal DF, Ohmann LF. Classification, description, and dynamics of upland plant communities within a Minnesota wilderness area. Ecological Monographs. 1975;**45**:389-407

[7] Heinselman M. Landscape evolution, peatland types, and the environment in the Lake Agassiz peatlands natural area, Minnesota. Ecological Monographs. 1970;**40**:235-261

[8] Dean JL. Wetland Forest Communities of the Eastern Boundary Waters Canoe Area. University of Minnesota; 1971. M.S. thesis. (Cited from Heinselmann, 1996)

[9] Joosten H. The Global Peatland CO_2 Picture: Peatland Status and Drainage Related Emissions in all Countries of the World. Ede: Wetlands International; 2009. Available from: https://unfccc.int/sites/ default/files/ draftpeatlandco2report.pdf

[10] Scharlemann JP, Tanner EV, Hiederer RR, Kapos V. Global soil carbon: Understanding and managing the largest terrestrial carbon pool. Carbon Management. 2014;**5**(1):81-91. DOI: 10.4155/cmt.13.77

[11] United Nations Environment Programme. 2022. Global Peatlands Assessment: The State of the World's Peatlands

[12] International Union for Conservation of Nature. Securing the Future for Global Peatlands. Hawaii: World Conservation Congress Resolution WCC-2016-Res-043-EN; 2016. Available from: https://portals.iucn.org/library/sites/library/filesresrecfiles/WCC_2016_RES_043_EN.pdf

[13] Burris JM, Haney AW. Bird communities after blowdown in a late-successional Great Lakes spruce-fir forest. The Wilson Bulletin. 2005;**117**(4):341-352

[14] Burris JM, Haney AW. Bird communities before and after a catastrophic blowdown in a Great Lakes pine forest. Bird Populations. 2006;7:10-20

[15] Lain EJ, Haney A, Burris JM, Burton J. Response of vegetation and birds to severe wind disturbance and salvage logging in a southern boreal forest. Forest Ecology and Management. 2008;**256**(5):863-871

[16] Apfelbaum SI, Haney A. Bird populations before and after wildfire in a Great Lakes pine forest. The Condor. 1981;**83**(347):354

[17] Apfelbaum SI, Haney A. Nesting and foraging activity of the Brown creeper in Northeastern Minnesota. The Loon. 1977;**49**(2):78-80

[18] Apfelbaum SI, Haney A. Changes in bird populations during succession following fire in the northern Great Lakes wilderness. In: Proceedings – National Wilderness Research Conference: Current Research, 10-16. General Technical Report INT GTR-212. Forest Service: Intermountain Research Station: U.S. Department of Agriculture; 1985;**6**

[19] Apfelbaum SI, Haney A. Bird population changes in recovering conifer forests documented (Minnesota and Ontario). Restoration and Management Notes. 1985;**3**(1):34-35

[20] Haney A, Apfelbaum S, Burris JM. Thirty years of post-fire succession in a southern boreal forest bird community. American Midland Naturalist. 2008;**159**(2):421-433

[21] Apfelbaum SI, Seelbach P. Nest tree, habitat selection, and productivity of seven north American raptor species based on the Cornell University Nest record card program. Raptor Research. 1983;**17**(4):97-113

[22] Apfelbaum SI, Haney A. Note of foraging and nesting habits of goshawks. The Loon. 1984;**56**(132):133

[23] ELPC, (Environmental Law and Policy Center). An Assessment of the Impacts of Climate Change on the Great Lakes2019. p. 70

[24] USGCRP, (US Global Climate Change Research Program). 4th National Review and Report. 1800 G Street, NW, Suite 9100, Washington, DC, 20006, USA; 2017. Available from: Globalchange.gov

[25] Byun K, Hamlet AF. Projected changes in future climate over the Midwest and Great Lakes region using downscaled CMIP5 ensembles. International Journal of Climatology. 2018;**38**:e531-e553

[26] Baker DG, Hines DA, Strub JH Jr. Precipitation facts, normal, and extremes. Climate of Minnesota, Part V, University of Minnesota Agricultural Experiment Station Tech. Bull. 1967:254

[27] Swain AM. A history of fire and vegetation in northeastern Minnesota as recorded in lake sediments. Quaternary Research. 1973;**3**:383-396

[28] Heinselman M. Fire and succession in the conifer forests of northern North America. In: West DC, Shugart HH, Botkin DB, editors. Forest Succession, Concepts, and Applications. New York, N.Y: Springer-Verlag; 1981. pp. 374-405

[29] Heinselman M. Fire intensity and frequency as factors in the distribution and structures of northern ecosystems. 1981:7-57. in H.A. Mooney, T.M. Bonnicksen N. L. Christensen, J.E. Lotan, and W.A. Reiners, coordinators Fire regimes and ecosystem properties. Proc. Of conference Honolulu, Hawaii, 1978. USDA Forest Service Gen. Tech. Rept. WO-26

[30] Natural Resources Canada. Canadian Forest Service. Frontline Express: Great Lakes Forestry, Bulletin #84

[31] Intergovernmental Panel on Climate Change. In: Hiraishi T, Krug T, Tanabe K, Srivastava N, Baasansuren N, Fukuda M, et al., editors. Supplement to the 2006 IPCC Guidelines for National Greenhouse gas Inventories: Wetlands: Methodological Guidance on Lands with Wet and Drained Soils, and Constructed Wetlands for Wastewater Treatment. Geneva, Switzerland; 2014. Available from: https://www.ipcc.ch/publication/2013-supplement-to-the-2006- IPCC-guidelines-for-national-greenhouse-gas-inventories wetlands/

Section 2

Evaluation of Mountain Ecosystems Response to Global Climate Change

Chapter 3

Forest Degradation in Dryland Ecosystems of Sudan: Review of the Causes, Consequences, Assessment Methods, and Potential Solutions

Emad H.E. Yasin, Ahmed A.H. Siddig, Eiman E. Deiab, Czimber Kornel, Ahmed Hasoba and Abubakr Osman

Abstract

Dryland forests are ecologically and socioeconomically important. They contribute to livelihood diversification, food security, animal feed and shelter, and environmental conservation in sub-Saharan Africa, particularly Sudan. Despite their importance, current findings show that multiple ecological, human, socio-economic, and policy factors have damaged these resources. As a result, undesirable consequences have been observed, such as food famine, land and water resource degradation, decline/loss of biodiversity, and contribution to global warming that affect the welfare of humans, plants, animals, and micro-organisms. This chapter briefly reviews the forest degradation in drylands Sudan with emphasis on its common causes, impacts, assessment methods, management intervention efforts, and potential future solutions. Given the current situation, there must be urgent combating efforts to manage Sudan's dryland forest resources properly. On the one hand, following prevention measures to essentially deal with the current causes thus prevent any further degradation of forest resources in dryland Sudan. On the other hand, there is an urgent need to address current degradation following appropriate and timely rehabilitation interventions. We also recommend adopting a serious monitoring and evaluation system within these combating efforts by applying the five common indicators for measuring forest degradation: biodiversity, productive functions, carbon storage, forest health, and protective functions.

Keywords: forest degradation, dryland ecosystems, causes of degradation, assessment methods & indicators, management interventions

1. Introduction

Forest resources in drylands hold significant socio-economic and ecological value, especially in sub-Saharan countries like Sudan. Here, they play vital roles in livelihood diversification, food security, animal feed, shelter, and environmental conservation. Despite their importance, recent studies reveal that these resources are undergoing severe degradation due to a complex interplay of natural, anthropogenic, socio-economic, and policy-related factors [1–3]. Forest degradation is a global issue that contributes to deforestation, desertification, and loss of biodiversity, particularly in drylands [4, 5]. Estimates suggest that approximately 850 million hectares of forests and forest lands are degraded [6, 7]. Drylands, which occupy 41% of Earth's land area, have already seen 10–20% degradation, with 1–6% of their populations living in desertified areas [8, 9]. Dry forests constitute more than 40% of all tropical forests [10, 11]. In Africa and tropical islands worldwide, dry forests account for 70–80% of forested areas. These ecosystems are characterized by their seasonality, differing significantly from rainforests, which remain stable year-round [12]. Due to their accessibility and fertility, dry forest zones have historically been preferred for human settlement over wetter forest zones [10–13]. However, this has led to extensive exploitation of these forests for thousands of years, including for fuelwood and charcoal production [14, 15]. In Sudan, forest resources are inversely proportional to population density. A staggering 68% of Sudan's forests are located in the south, where only 15% of the population resides. In contrast, the more populated northern states hold just 32% of the forests [16–19]. The 2015 National Biodiversity Strategy and Action Plan (NBSAP) showed that forest cover accounts for a mere 11.9% of the country, with an annual deforestation rate exceeding 2.4%. The dry tropics are characterized by shifting cultivation systems with both long cultivation and short fallow periods, which hamper the natural recovery of vegetation [20, 21]. Despite their long history of exploitation and the associated degradation, dry forests have received scant attention compared to rain forests. Knowledge gaps make the management and restoration of these forests a challenge, with little data available on forestry activities, stock assessment, and the economic contributions of these resources [20, 21]. In this review, we shed light on the issue of forest degradation in Sudan's drylands. We focus on internationally recognized indicators of forest degradation, its causes, and its socio-economic and environmental consequences. We also propose a set of recommendations for preventing, controlling, and reversing forest degradation as a direction for future research and policy.

2. Concept and signs of forest degradation

Teketay [22] defines forest degradation as alterations within a forest that negatively impact its structure and function, thereby reducing its ability to supply products and/or services. In practical terms, however, the concept of degradation is far more subjective, as perceptions can vary widely regarding the same landscape. Various definitions of forest degradation have been examined through literature reviews, and these are summarized below. The criteria used to define forest degradation differ among authors and international bodies, as highlighted by [23]. This divergence in definitions further emphasizes the subjectivity inherent in understanding and assessing forest degradation, making it a complex and nuanced issue that requires comprehensive, multi-dimensional approaches for effective management and mitigation (**Table 1**).

Overall, a common thread in most definitions of forest degradation found in the literature is the emphasis on reduced crown cover and loss of biodiversity as key indicators. These two elements often serve as crucial signs pointing to the declining health and function of a forest, thus making them vital parameters for evaluating the extent and severity of forest degradation. This consensus provides a starting point for

Parameter	References covering the parameters
Changes within the forest	
Structure	[6, 17, 23–27]
Crown cover	[14, 16, 23, 28–32]
Species composition	[12, 16, 17, 23, 26]
Biodiversity	[12, 17, 23, 25]
Stocking	[23, 25, 26]
Soil degradation	[23, 26]
Trees density	[27]
Forest health	[23]
Reduction of capacity to provide	
Productivity	[6, 22, 23, 27, 32–35]
Goods	[6, 23, 25, 26, 28]
Services	[6, 23, 25, 26, 28]
Wood	[6, 25].
Carbon stocks/biomass	[23, 26, 29, 36, 37].
Other functions	[12, 23, 25, 26, 36]

Sources: Adopted from [23].

Table 1.
Parameters covered by different definitions of forest degradation from the literature review.

a more unified approach to monitoring and mitigating the issue, despite the various nuances and subjectivities involved in defining forest degradation.

3. Causes of drylands' forest degradation

Despite the immense value and services that forest ecosystems offer in drylands, these vital resources face numerous threats that contribute to their degradation. The literature has extensively discussed the types and categories of factors leading to forest degradation. Generally, there seems to be a consensus to classify these factors into natural and anthropogenic categories [38].

Additionally, the causes of dryland forest degradation (DFD) span multiple dimensions, including economic, social, ecological, policy, and governance factors. These causes can further be categorized under natural influences as well as human-induced (anthropogenic) and social and policy-related factors. This multi-faceted nature of the drivers behind DFD underscores the need for an integrated approach that addresses each of these dimensions to effectively manage and mitigate the degradation of these critical ecosystems.

3.1 Natural factors of forest degradation

Natural factors contributing to dryland forest degradation (DFD) can be further subdivided into:

- *Physical environmental factors*: These include bushfires, challenging topographical conditions such as sloping terrain, and erosion caused by both wind and water. Soil fertility decline also plays a part, often associated with physical problems like low organic matter content, salinity, and alkalinity.

- *Climatic factors*: Insufficient and variable rainfall, unpredictable variations in rainfall patterns both within and between seasons, and intermittent but severe drought conditions all adversely affect the natural regeneration of both indigenous and plantation forests.

- *Biological factors*: These encompass diseases and pest infestations, which may lead to forest fires in extreme cases. For instance, pest outbreaks in lowlands can result in the burning of forests and woodlands, thereby constraining tree-planting efforts. Additionally, aggressive invasive perennial plants and termite attacks can significantly harm the forest ecosystems.

Each of these categories poses unique challenges for the preservation and restoration of dryland forests, and understanding them is crucial for the formulation of effective management and mitigation strategies.

3.2 Anthropogenic factors of forest degradation

Major anthropogenic factors contributing to dryland forest degradation include:

- *Deforestation*: This involves the destruction, clearing, or incineration of dryland forests and woodlands for several reasons: a) Expansion of crop cultivation, driven by a rising human population and limited economic diversification options. b) Production of charcoal and wood for domestic energy needs, due to a lack or shortage of alternative energy sources, as well as for construction materials and carpentry items. These resources are also often sold to generate income and support household livelihoods. c) Immediate socio-economic and infrastructural development needs, such as resettlement, mining, and road construction.

- *Overgrazing/grazing by livestock*: Overgrazing leads to soil compaction and damage to seedlings, hampering natural regeneration processes.

- *Unsustainable utilization*: This includes improper and unplanned harvesting practices, which result in significant wood wastage due to extremely low recovery rates and damage to the residual trees, plants, and stands.

- *Introduction of invasive alien species*: The introduction of non-native species can lead to the displacement of native dryland forest flora and, by extension, the reduction in overall biodiversity.

Each of these anthropogenic factors poses its own set of challenges to the health and sustainability of dryland forests. Addressing them will require multi-faceted and tailored approaches that take into account the complexity and interconnectedness of these contributing elements.

4. Socio-economic and policy-related factors

According to Teketay [22], several socio-economic and policy-related factors contribute either directly or indirectly to the degradation of dryland forests. The major factors among these include:

- *Poverty, population growth, and poor economic performance*: These interconnected factors exacerbate the pressure on dryland forests as communities rely more heavily on forest resources for their livelihoods.

- *Inadequate or absence of land-use classification and forest policies*: The lack of clear land-use classification, along with insufficient or absent forest policies and legislation, leaves forests vulnerable to exploitation and degradation.

- *Absence of land and tree tenure/ownership rights*: Without clear rights to land or trees, there's little incentive for individuals to manage forests sustainably.

- *Underestimation of dryland forests' contribution*: The economic and ecological value of dryland forests often goes unrecognized, leading to policy neglect.

- *Lack of pricing and incentive policies*: Without economic incentives or appropriate pricing mechanisms, sustainable forest management practices are rarely adopted.

- *Inadequate institutional arrangements*: Weak organizational structures and governance mechanisms can hinder effective forest management.

- *Absence or inadequacy of viable dryland forestry development strategies*: The lack of long-term planning and investment in sustainable forestry initiatives contributes to ongoing degradation.

- *Weak forestry research systems*: Insufficient research into sustainable forestry practices limits the development of effective forest management techniques.

- *Insufficient information management*: A lack of comprehensive and accessible data hampers the planning and execution of sustainable forestry initiatives.

- *Political instability*: Civil wars, unrest, and other forms of political instability can result in short-term exploitation of forests at the expense of long-term sustainability.

Understanding and addressing these socio-economic and policy-related factors is crucial for developing multi-pronged strategies to manage and mitigate dryland forest degradation effectively.

5. Consequences of dryland forest degradation

The impacts of dryland forest degradation are multi-dimensional and significantly detrimental to both the environment and human livelihoods. Here are some key consequences:

1. **Environmental consequences:**

    - *Wood famine:* Over-exploitation of forest resources leads to a wood deficit, which can prompt communities to "mine" forests, i.e., to harvest wood at rates exceeding the forest's capacity for natural regeneration [22].

    - *Land degradation:* This refers to both soil erosion and loss of soil fertility. Often, metrics used to assess land degradation focus on the severity of soil erosion, which can be particularly acute in areas with high-intensity rainstorms and steep slopes [39, 40].

    - *Degradation of water resources:* Poor land management can lead to sedimentation and siltation in water bodies. Deforestation and soil erosion increase surface runoff, reducing the soil's capacity to absorb and store water, affecting groundwater levels and overall water availability [22, 41].

    - *Decline of biodiversity:* Reduced vegetation cover and invasion by alien species undermine the ecological health of dryland forests, leading to a decline in both flora and fauna. Loss of biodiversity affects valuable non-timber forest products (NTFPs), such as wild foods, medicine, and materials for crafts, that communities rely on [22, 42].

    - *Enhanced global warming:* Deforestation contributes to greenhouse gas emissions, exacerbating global warming. Additionally, forests act as carbon sinks, and their degradation means less carbon is being sequestered from the atmosphere [37, 43, 44].

2. **Socio-economic consequences:**

    - *Food and livelihood security:* The degradation of dryland forests threatens food security by reducing the availability of wild foods and other forest resources that rural communities rely on.

    - *Economic value:* Forests offer various forms of livelihood, from timber to non-timber forest products like resins, gums, and honey. Forest degradation diminishes these opportunities, impacting the local and national economies.

    - *Public health:* Poor water quality from sedimentation and reduced availability of medicinal plants can have implications for public health.

    - *Cultural and spiritual value:* Many communities have deep cultural and spiritual connections to forests, and degradation can result in the loss of these important aspects of human well-being.

To address these challenges, a multi-faceted approach is needed that combines sustainable forest management practices, socio-economic interventions, and policy changes. This should also involve the active participation of local communities and be guided by comprehensive research and monitoring (**Figure 1**) [45].

This diagram illustrates the complex interplay of processes contributing to forest degradation in the Brazilian Amazon. The framework begins with pristine forests that

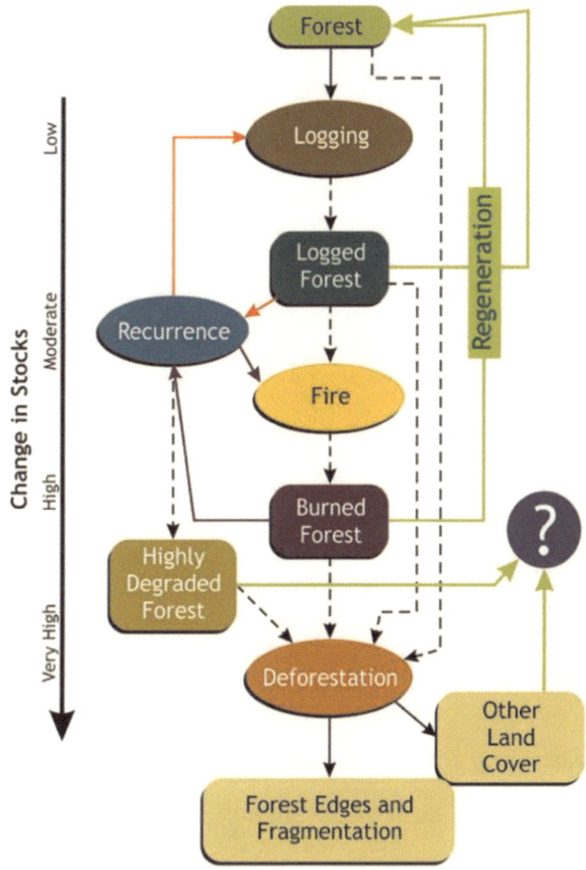

Figure 1.
Dynamics and interactions of forest degradation [45].

become vulnerable through several stages that are mentioned in the diagram and seeks to encapsulate the chain reactions and feedback loops involved in forest degradation, emphasizing the multifaceted nature of the issue.

Figure 2 presents a detailed look at the far-reaching implications of forest cover loss. At its core, the loss of forest cover serves as a catalyst for a cascade of negative environmental, social, and economic consequences. The environmental impacts are immediate and devastating, affecting biodiversity, soil quality, and the water cycle. This leads to a decrease in carbon sequestration capabilities, which further exacerbates climate change. On a social level, forest cover loss adversely affects indigenous communities and local populations that rely on forests for their livelihoods, leading to displacement and poverty. The economic repercussions are equally severe, with the loss of valuable resources like timber, non-timber forest products, and ecosystem services that have both local and global economic value. The figure also illustrates how these different domains are interlinked, emphasizing that the implications of forest cover loss are not isolated but are interconnected in complex ways. Therefore, forest cover loss represents a multi-dimensional issue that requires integrated solutions.

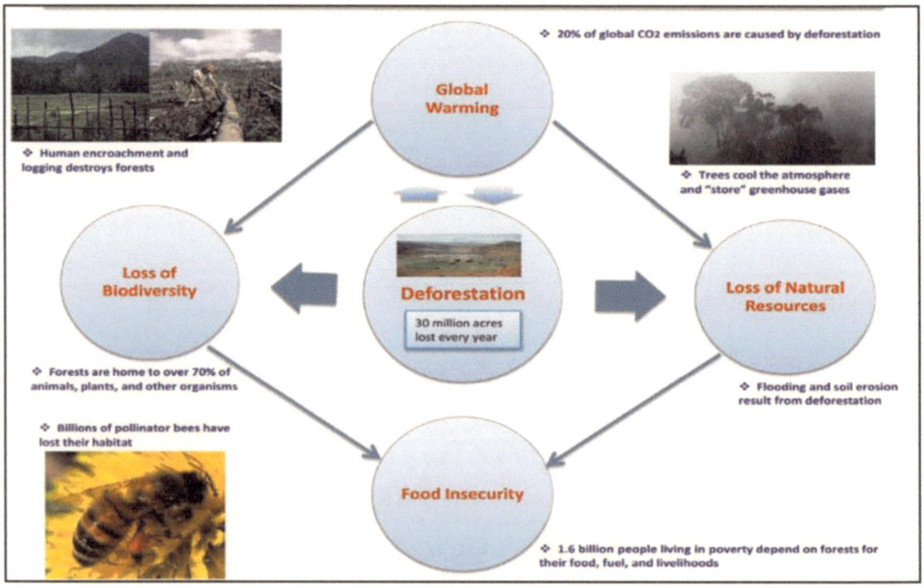

Figure 2.
Implication of forest cover loss [17].

6. Methods of assessing forest degradation and common indicators

There are two common methods for assessing forest conditions which are briefly stated in the following:

6.1 Remote sensing-based multi-temporal analysis

Advancements in modern technology have been a boon to the field of environmental science, particularly in the realm of forest conservation and management. Among these technologies, remote sensing stands out as a particularly effective method for long-term monitoring of forest conditions. Developed over the years through various studies [44, 46, 47], remote sensing technologies now allow for sophisticated multi-temporal analyses that can reveal subtle but significant changes in forest ecosystems.

In the multi-temporal analysis approach, satellite images of the same geographical location are captured at different times and are then aligned or 'co-registered' with one another. This enables scientists to directly compare these images, paying particular attention to changes in spectral values, which are indicative of shifts in vegetation condition and health. Spectral values are measures of how different wavelengths of light are absorbed or reflected by the Earth's surface, providing a sort of 'fingerprint' of its current state. For instance, healthy vegetation reflects a lot of green light but absorbs red and blue light, which is why it appears green to the human eye. By studying these spectral "signatures," experts can glean intricate details about soil moisture levels, plant health, and even species composition.

The utility of this approach has been demonstrated in various settings, including the ability to monitor specific types of environmental stress. For example, the analysis of multi-temporal imagery has proven to be particularly effective for tracking defoliation events caused by insect infestations [48]. Such defoliation would show up as changes in the spectral values, allowing researchers to not only identify affected areas quickly but also estimate the extent of foliage loss. This is invaluable information for forest management teams who can then take targeted action, whether it be through the introduction of natural insect predators or other integrated pest management strategies.

Moreover, multi-temporal remote sensing offers the advantage of continuous, automated monitoring, enabling timely interventions. It's an approach that makes it possible to identify patterns and trends over time, from seasonal variations to the long-term impacts of climate change. This not only facilitates proactive management practices but also aids in policy formulation for sustainable forest management.

While remote sensing provides a bird's-eye view of large forested areas and can quickly detect large-scale changes, it's important to combine this method with ground assessments for a more comprehensive understanding. On-the-ground assessments can validate satellite imagery findings while adding context, especially when it comes to local biodiversity, soil conditions, and human activity. Therefore, the integration of both remote sensing and ground-based methods offers the most thorough, accurate, and actionable insights for forest conservation efforts.

6.2 Ground assessment of vegetation conversion and change

In addition to the technological advances of remote sensing, direct field or ground assessments serve as another crucial method for monitoring and evaluating forest conditions. Unlike remote sensing, which offers a macroscopic, overhead perspective, ground assessments provide a microscopic, detailed view that is indispensable for a comprehensive understanding of forest ecosystems.

Ground assessments typically involve systematic vegetation sampling to evaluate key metrics such as species composition, abundance, distribution, and structural attributes of the forest. Essentially, researchers collect baseline data for these parameters and then periodically revisit the site to identify any changes. This approach is cited as particularly effective for assessing forest degradation, offering nuanced insights that remote sensing sometimes misses [45, 49, 50].

During these ground assessments, various indicators can be meticulously examined. For instance, vegetation composition can be assessed through species identification and counts, helping to determine if invasive species are encroaching upon native vegetation. Abundance and distribution metrics can indicate whether certain species are becoming more dominant or rare, possibly signaling ecological imbalance. Structural attributes, such as tree height, canopy density, and undergrowth conditions, can provide clues to the forest's overall health and its ability to support various forms of wildlife.

Importantly, ground assessments also allow for the study of soil conditions, local hydrology, and even microclimatic variables, offering a holistic perspective on the forest ecosystem. Soil samples can reveal nutrient deficiencies, contamination, or changes in microbial communities, all of which have far-reaching implications for vegetation health. Examination of local water sources can also identify any changes in water quality or quantity, which again impacts the entire ecosystem.

Additionally, ground assessments can be designed to be adaptive, allowing researchers to adjust their monitoring strategies based on preliminary findings or

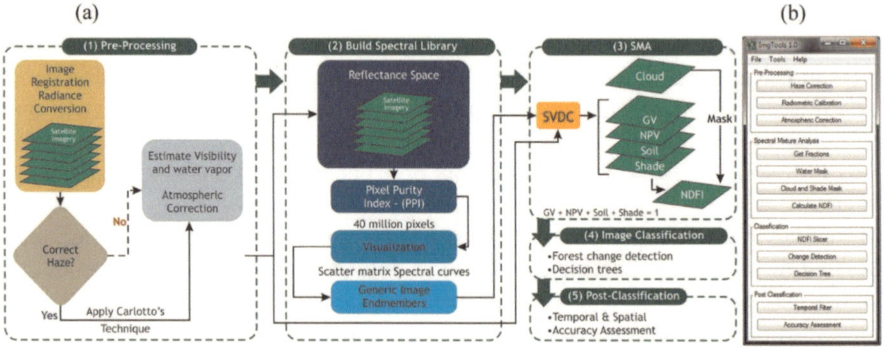

Figure 3.
Conceptual framework of the image processing procedures: a) pre-processing, b) software [45].

emerging environmental threats. This adaptability ensures that the assessment remains relevant and continues to generate actionable data over time.

While ground assessments are generally more labor-intensive and time-consuming than remote sensing methods, they offer an unparalleled level of detail. Therefore, the most effective forest monitoring programs often employ a multi-faceted approach, combining the broad overview provided by satellite imagery with the granular detail captured through ground assessments. This integrated methodology ensures a more accurate and nuanced understanding, which in turn facilitates better decision-making for conservation and management efforts (**Figure 3**).

6.3 Common indicators of forest degradation in dryland ecosystem

Forest degradation has been succinctly defined as a reduction in the ability of a given forest ecosystem to provide goods and services [51]. A comprehensive framework was first established in 2009 and has been globally adopted to measure forest degradation, including in dryland areas. This framework consists of five major categories of indicators:

1. *Biodiversity indicators*: These gauge the ecosystem's biological health and include measures like changes in ecosystem diversity, shifts in ecosystem resilience (or state), and metrics related to fragmentation, road density, and the abundance of indicator species. Additional aspects may involve monitoring keystone and flagship species, which can indicate the general state of biodiversity in a specific region.

2. *Productive functions indicators*: These are concerned with the forest's economic and resource potential. They measure changes in the amount of timber products, growing stock for selected species, non-wood forest products, fuelwood, and even water quality and quantity. This also extends to measuring the impact of forest management practices on the productivity of these resources over time.

3. *Carbon indicators*: With a focus on climate change and carbon sequestration, these indicators monitor the amount of total biomass, growing stock, and carbon stored in each of the five carbon pools (above-ground biomass, below-ground biomass, litter, deadwood, and soil organic carbon). They may also incorporate

additional measures like changes in leaf area index, which can provide insights into the forest's ability to sequester carbon.

4. *Forest health indicators*: This category addresses the forest's vulnerability to both biotic (living) and abiotic (non-living) agents, such as fires, diseases, invasive species, storms, and snowfall. Sub-indicators include:

6.3.1 Crown condition

The crown indicator serves as a comprehensive measure to evaluate the health and vitality of individual trees within a forest ecosystem. It specifically focuses on three main components: the amount of foliage, the condition of that foliage along with branches, and the distribution of growing tips on the trees. By assessing these components, one can derive valuable insights into not only the health of individual trees but also the overall health and resilience of the forest.

1. *Amount of foliage*: The density and amount of leaves or needles on a tree can provide essential information about its nutritional status, potential for photosynthesis, and by extension, its general health. Sparse or discolored foliage may indicate nutritional deficiencies, disease, or environmental stressors affecting the tree.

2. *Condition of foliage and branches*: The physical state of the leaves and branches, including signs of disease, insect damage, or physical injury, can reveal the immediate challenges facing a tree. This could range from pests and pathogens to environmental stressors like drought or pollution.

3. *Distribution of growing tips*: The presence and distribution of growing tips, or meristems, can offer insights into the tree's reproductive potential and future growth. An imbalance in the distribution might suggest that the tree is experiencing stress or has been affected by external factors like pruning or damage.

   Together, these aspects offer a nuanced picture of trees and, by extension, forest health. They can be invaluable for early detection of forest degradation factors such as disease outbreaks or the impact of environmental stressors. Given their significance, crown indicators have been elaborated extensively in academic and practical literature, serving as a cornerstone in the study and monitoring of forest ecosystems [52].

6.3.2 Lichen communities

Lichens, the symbiotic organisms resulting from the partnership between fungi and algae, are a remarkable component of many forest ecosystems. They colonize various surfaces, including the bark of trees, rocks, and even soil, thereby playing multiple crucial roles within these ecosystems.

1. *Nutrient cycling*: Lichens are particularly effective at breaking down rocks and other hard substrates, contributing to the formation of soil and the cycling of essential nutrients. This process is critical for the ongoing productivity and health of forest ecosystems.

2. *Wildlife habitat and forage*: Numerous small animals and insects rely on lichens for both habitat and food. For example, lichens serve as nesting materials for birds and offer a vital winter food source for some species of herbivores.

3. *Environmental indicators*: Lichens are highly sensitive to various environmental factors, including the quality of the air. Their physiology allows them to absorb nutrients and moisture directly from the atmosphere, which also makes them vulnerable to pollutants. Therefore, changes in lichen communities can be an early warning sign of deteriorating air quality or other ecological stresses.

4. *Climate change indicators*: Lichens have also been found to be highly responsive to changes in climate, including temperature and humidity variations. Their presence or absence can serve as a valuable barometer for assessing long-term climate trends and their impact on forest ecosystems.

   Given their multi-faceted contributions and sensitivity to environmental conditions, lichens serve as an invaluable bio-indicator for forest health and vitality. Monitoring the diversity, abundance, and condition of lichens can provide a wealth of information about the forest's current state and its resilience to future changes or threats. This concept of using lichens as an indicator has been described extensively, providing a foundational tool for ecologists, conservationists, and researchers interested in forest health and sustainability [53].

6.3.3 Down woody debris

Down Woody Debris (DWD) serves as an often-overlooked yet invaluable metric for assessing the health, complexity, and functionality of a forest ecosystem. This indicator involves the meticulous cataloging and analysis of various forms of woody material that have fallen to the forest floor, ranging from smaller twigs to entire tree trunks.

1. *Classification by size*: DWD is typically divided into two main categories based on size. Smaller fragments with diameters less than 3 inches are classified as "fine woody debris," while larger branches and entire fallen trees are termed "coarse woody debris."

2. *Species identification*: Documenting the species of the downed woody material can offer critical insights into the forest's biological diversity and the specific types of trees that may be more susceptible to disease, pests, or environmental stresses.

3. *Shape, size, and decay*: The varying shapes, sizes, and stages of decay are also recorded. These characteristics not only help in assessing the rate of decomposition and nutrient cycling but also offer clues about the forest's history and the ecological processes at play.

4. *Habitat and ecological importance*: DWD serves multiple ecological roles. It provides habitat for a range of organisms, from fungi and insects to larger mammals. It's also a critical component in the forest's nutrient cycle, as it decomposes to release essential minerals back into the soil.

5. *Monitoring decay and environmental change*: The rate of decay can reveal information about climatic conditions such as moisture levels and temperature, making DWD a useful proxy for monitoring environmental change over time.

6. *Safety and fire risk*: Excessive DWD can pose fire risks in certain conditions. Therefore, monitoring DWD levels can be essential for forest management, especially in regions prone to wildfires.

7. *Data-driven forest management*: Accurate DWD measurements contribute to more effective forest management practices, helping resource managers make informed decisions regarding activities like controlled burns, logging, and conservation efforts.

   By meticulously examining and cataloging Down Woody Debris, researchers and forest managers can glean a multitude of insights about the ecosystem's health, its history, and its capacity to support various forms of life. These aspects of the DWD indicator have been discussed in great depth, offering a nuanced tool for understanding and managing forest ecosystems [54].

6.3.4 Tree damage

The Damage Indicator serves as a comprehensive gauge for evaluating the various impacts affecting the health and vitality of a forest ecosystem. This measure delves into the nuances of forest health by meticulously recording different types of injuries and their causative agents. By doing so, it allows for a more granular understanding of forest health, thereby aiding in targeted conservation and management efforts.

1. *Types of damages*: The indicator categorizes damages into multiple types, ranging from diseases and pest infestations to weather events like storms and human-induced injuries such as logging or accidental fires.

2. *Severity and extent*: It's not just about identifying the types of damage but also evaluating their severity and extent. For instance, the spread of a specific canker disease through a tree stand may be mapped and classified based on how debilitating it is to the trees.

3. *Localization*: Pinpointing the exact location of the damage within the forest and even within individual trees (e.g., trunk, roots, or branches) offers more actionable data for remedial measures.

4. *Recording specific ailments*: Data collection extends to specific symptoms such as open wounds, signs of advanced decay, cankers, and structural damages like broken boles and roots. Each of these parameters can indicate a different set of environmental stressors or diseases affecting the forest.

5. *Monitoring over time*: Tracking these damage metrics over a period allows forest managers to spot trends or spikes in forest health issues, enabling timely interventions.

6. *Impact on ecosystem services*: Understanding the types and extents of damages is crucial for assessing the forest's capacity to continue providing ecosystem services, whether they are in the form of timber, non-wood forest products, or even intangibles like air and water purification.

7. *Data-driven decision making*: By compiling this intricate data, forest managers and policymakers can make more informed decisions, whether it's implementing disease control measures, modifying logging practices, or planning restoration activities.

8. *Emergency response*: In cases of sudden and severe damage, like that caused by storms or pest outbreaks, this indicator can be crucial for orchestrating rapid and effective response strategies.

9. *Public awareness and education*: Publishing findings related to the Damage Indicator can also serve an educational role, informing the public about the state of their local forests and the importance of responsible stewardship.

   By methodically recording and interpreting these aspects of tree damage, this indicator furnishes a multi-faceted lens through which forest health can be evaluated and managed [55].

6.3.5 Tree mortality

The Tree Mortality Indicator is a crucial metric that offers insights into the overall health and sustainability of a forest ecosystem. It meticulously quantifies the number, size, and volume of trees that have died, thereby furnishing critical data that can illuminate the underlying causes of forest decline. In doing so, it allows for nuanced approaches to forest management and conservation.

1. *Number of trees*: The indicator starts by counting the actual number of trees that have succumbed to death. This count is valuable in assessing the scale of tree mortality within a specific forest or region.

2. *Size and volume*: In addition to the sheer number, the size (diameter at breast height, DBH) and volume of the dead trees provide insights into which age classes or species are more susceptible to certain threats. For example, older trees may be more vulnerable to diseases, while younger ones might be more susceptible to drought.

3. *Categorizing stressors*: The indicator helps identify whether the cause of tree mortality is related to biotic factors like diseases and pests or abiotic factors like drought, fire, or human activity. This categorization aids in devising specific interventions to mitigate future losses.

4. *Stand development and structural changes*: Changes in forest stand composition and structure can lead to altered growing conditions that may become unfavorable for specific tree species or age classes. Monitoring this data can help in forest succession planning.

5. *Indicator of ecosystem health*: High tree mortality rates can signal broader ecosystem instability, affecting not just the flora but also the fauna that depend on them.

6. *Impacts on carbon sequestration*: Tree mortality directly affects the forest's ability to act as a carbon sink. Dead trees release stored carbon back into the atmosphere, thus playing a role in larger climate dynamics.

7. *Adaptive management*: Information from the Tree Mortality Indicator can be used for adaptive forest management strategies. For example, if a certain tree species shows high mortality rates, forest managers may consider planting more resilient species.

8. *Policy implications*: High levels of tree mortality could prompt regulatory changes, such as altering logging quotas or implementing more stringent conservation policies.

9. *Long-term monitoring*: Continual monitoring allows for the tracking of tree mortality trends over time, helping to validate the effectiveness of implemented interventions or the need for new strategies.

10. *Public engagement*: Transparently sharing these mortality statistics can raise public awareness about the health of local forests, thus fostering community engagement in conservation efforts.

Overall, the Tree Mortality Indicator serves as a robust diagnostic tool that helps stakeholders better understand forest dynamics, thereby allowing for more effective and targeted conservation initiatives [56].

6.3.6 Vegetation diversity and structure

The vegetation diversity indicator provides a comprehensive evaluation of forest ecosystems by quantifying the variety, abundance, and spatial arrangement of vascular plant species. This tool serves multiple purposes, including inventory assessment and long-term monitoring. By keeping track of changes in species richness, their relative abundances, and the layering or vertical structure within the forest, it helps researchers and conservationists gauge the health and resilience of the forest over time. This indicator is particularly useful for capturing nuanced shifts in forest composition, which can result from various factors like climate change, habitat fragmentation, or human activity. Additionally, changes in the vertical stratification of vegetation—such as the understory, midstory, and canopy—can indicate the forest's capacity to support a diverse array of wildlife and maintain its ecological functions. This comprehensive approach to understanding vegetation diversity is invaluable for adaptive management strategies, providing a more complete picture of forest health and functioning. For a more in-depth discussion and methodology of this indicator [57].

6.3.7 Soil condition

Soil serves as the foundational component in forest ecosystems, supplying the essential environmental elements upon which all vegetation relies: nutrients for growth, water for sustenance, air for respiration, heat for temperature regulation, and mechanical support for stability. Beyond these basic functions, soil health is intricately linked to a range of forest processes and services, thereby making it a crucial indicator for forest management and conservation. The soil condition indicator provides both chemical and physical data, offering critical insights into the soil's fertility status and its physical characteristics such as texture, structure, and water-holding capacity. These metrics are instrumental in a variety of ecological models, which operate at diverse spatial scales within the forest ecosystem. For instance, accurate soil data can help estimate the forest's carbon budget, thereby assessing its role in carbon sequestration and climate change mitigation. By comprehensively understanding the condition of the soil, researchers and forest managers can make more informed decisions about sustainable land use, reforestation initiatives, and the mitigation of environmental impacts such as soil erosion and nutrient depletion. This deep-dive into soil indicators and their significant role in forest ecology is elaborated upon in the study detailed by [58].

6.3.8 Ozone injury

A subset of plant species serve as particularly sensitive bioindicators of ozone exposure levels that exceed typical background concentrations. These plants display distinct symptoms of foliar injury on the upper surface of their leaves, which sets them apart from other types of foliage damage. Such easily identifiable and quantifiable signs of ozone-induced stress make them invaluable for monitoring air quality and its potential impact on forest ecosystems. By focusing on these bioindicator species, researchers can more precisely gauge the levels of ozone pollution in a given area and understand its ecological implications. This information is especially critical for assessing the health of forest ecosystems that may be affected by rising ozone levels due to industrial activities or other environmental stressors. The presence or absence of these symptoms can serve as a red flag for policymakers and forest managers, enabling them to take timely action to mitigate harmful effects. The intricate dynamics of how ozone levels influence plant health, and by extension, the overall well-being of forest ecosystems, is explored comprehensively in the research elaborated upon by [59].

6.3.9 Protective function indicator which is indicated by the rate of soil erosion (or area affected)

The Protective Function Indicator focuses on assessing the forest's role in soil conservation, particularly by monitoring the rate of soil erosion or the extent of areas affected by it. Soil erosion can lead to the loss of fertile topsoil and decrease the forest's ability to store water and nutrients, adversely impacting its overall health and resilience.

This indicator is crucial because forests often serve as natural barriers against soil erosion, thereby preserving water quality and reducing the risk of landslides and other natural disasters. High rates of soil erosion could signify that the forest's protective functions are compromised, possibly due to factors such as deforestation, uncontrolled logging, or the presence of invasive species that destabilize soil structures.

By closely monitoring the rate of soil erosion and the extent of affected areas, forest managers can take targeted conservation measures, such as reforestation or implementing erosion control structures like terraces or sediment traps.

Understanding the forest's ability or inability to perform this protective function can be pivotal for land-use planning and natural resource management strategies.

The methodology and metrics for assessing the Protective Function Indicator in the context of soil erosion are discussed in detail in the relevant scientific literature, serving as a guideline for both researchers and policymakers in the evaluation of forest degradation and conservation efforts.

7. Prevention and control measures of forest degradation in drylands

Addressing the pervasive issue of forest degradation in drylands necessitates an intensive, multi-faceted approach, given the magnitude and far-reaching implications of the problem. However, merely dealing with the degradation after it occurs is insufficient; it's vital to take into account the intricate web of contributing factors—be they climatic, anthropogenic, or ecological—that feed into this environmental calamity. This involves not just identifying its root causes, but also establishing a comprehensive set of assessment indicators that can measure the degradation's scope, rate, and ultimate impact on both the local ecosystem and broader environmental health.

Beyond immediate control and rehabilitation methods, a proactive, preventive strategy must also be part of the solution. Simply put, measures to halt degradation are not enough; preemptive actions that thwart its onset in the first place are just as crucial. This calls for a shift in perspective, from a reactive to a proactive stance in combating forest degradation.

With this in mind, we propose two key interventions aimed at both effectively halting and preventing forest degradation in drylands:

1. Comprehensive Ecosystem Monitoring: Implement a real-time, data-driven monitoring system that uses advanced technologies like remote sensing, GIS mapping, and predictive modeling. This would not only provide a detailed understanding of the current state of forest health but would also allow for the early detection of signs of degradation. The sooner these signs are detected, the quicker and more effective preventive measures can be taken.

2. Sustainable land management: Introduce land-use policies that are ecologically sustainable, such as agroforestry and controlled grazing, and combine these with community education and involvement. By adopting such integrated approaches, not only can immediate threats be managed, but the local community can also be empowered to act as stewards of their environment, thus ensuring long-term sustainability

7.1 Prevention/mitigation measures:

Addressing the root causes of forest degradation is an intricate task that involves a multi-pronged approach, taking into account not just the environmental aspects but also forest governance, community involvement, and economic factors. This complexity demands a multi-layered solution tailored to specific concerns.

1. **Forest governance:**

    - *Establishing a professional body for foresters*: A dedicated organization can provide guidelines, share best practices, and set ethical standards. This body can serve as a liaison between governmental agencies and local communities.

- *Harmonizing laws*: Streamlining forestry laws can prevent conflicting rules that may result in ineffective management or misunderstandings.
- *Awareness programs*: Conducting educational campaigns to raise awareness about forestry laws can help avoid conflicts and violations.
- *Local NGO involvement*: Encouraging NGOs to incorporate forest conservation programs can augment government efforts in preventing degradation.

2. **Community involvement**:

- *Community-led forest management*: Empowering local communities to take part in the management of forests not only promotes conservation but can also provide local economic benefits.
- *Participative monitoring*: Local communities, being the closest to the forest resources, can be vital in monitoring and reporting signs of degradation or illegal activities.

3. **Economic factors**:

- *Poverty alleviation programs*: Implementing social safety nets and income-generating opportunities can reduce the dependency of families on forest resources for their livelihoods.
- *Alternative energy sources*: Providing affordable and clean alternative energy options can significantly reduce the unsustainable use of forest resources for fuel.
- *Regulating timber trade*: A comprehensive system should be in place to combat illegal timber trade and regulate its export.
- *Recognizing traditional medicine*: Understanding and officially recognizing the value of forest plants for traditional medicine can make sustainable harvesting more economically viable.
- *Modern agricultural systems*: Implementing modern, sustainable agricultural practices can reduce the pressure exerted on forest land for agriculture.

By focusing on these pillars—governance, community involvement, and economic factors—tailored interventions can be developed to tackle the root causes of forest degradation. Doing so involves not just policy changes but also active collaboration among governments, professionals, communities, and NGOs.

7.2 Control/adaptation measures

To holistically address the challenges of forest degradation in drylands, a multifaceted approach is necessary. Here are some expanded guidelines grouped under specific topics that could pave the way for more sustainable forest management:

1. **National forest programs and eco-agriculture**:

    - *Comprehensive national forest programs*: Creating realistic, yet ambitious, national frameworks can provide a roadmap for long-term forest conservation and restoration.

    - *Introduce eco-agriculture*: Adopting agroecological methods that integrate trees and tree products can offer dual benefits: enhanced agricultural yield and forest conservation.

    - *Develop markets and infrastructure*: Building and standardizing marketplaces for tree products encourages sustainable usage and boosts local economies.

2. **Market exploration and price regulation**:

    - *Explore both domestic and international markets*: Widen the market access for dryland forest products to optimize revenue generation.

    - *Timely market information*: Providing real-time market data can help producers understand demand and get fair prices.

    - *Price regulation*: Ensuring price floors for dryland forest products can make forest conservation more economically viable.

3. **Alternative energy sources**:

    - *Promote non-wood energy*: Encourage the use of energy alternatives like solar, wind, and biogas to reduce dependency on wood, charcoal, and other forest resources.

4. **Integrated and multidisciplinary approach**:

    - *Comprehensive solutions*: Given the complexity of dryland forest degradation, an integrated approach that combines sectoral development with targeted interventions is vital.

5. **Information dissemination and research**:

    - *Organize and publish grey literature*: The knowledge that has been accumulating but remains unpublished should be organized, published, and disseminated for wider benefit.

    - *Valuation of dryland forest products*: Conduct studies to place monetary values on the ecosystem services provided by dryland forests.

    - *Cross-disciplinary research*: Investigate not just the biophysical factors but also the social, economic, and institutional aspects of forest degradation and rehabilitation.

By methodically tackling each of these areas, it is possible to build a comprehensive strategy that not only mitigates the ongoing degradation but also sets a foundation for the sustainable development of dryland forests [60].

Figure 4 presents a comprehensive conceptual framework that delineates the intricate relationship between the causes and consequences of land degradation, as

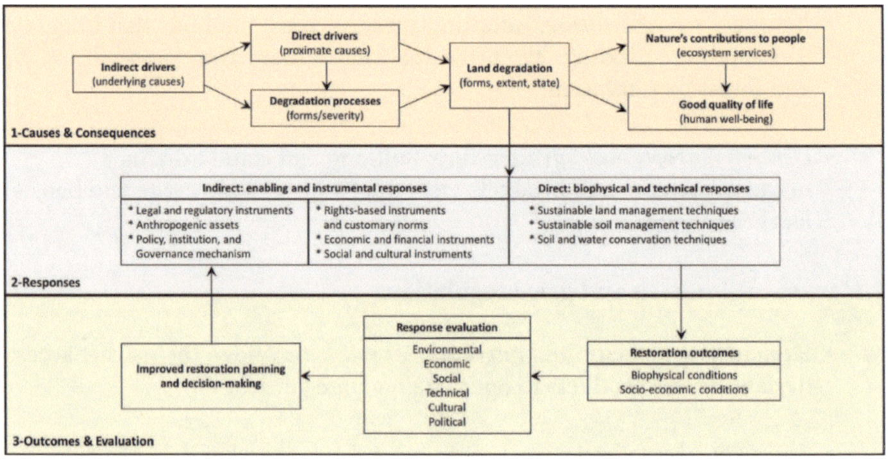

Figure 4.
Conceptual framework linking land degradation causes and consequences, restoration responses (both indirect and direct), and response outcomes and evaluation [60].

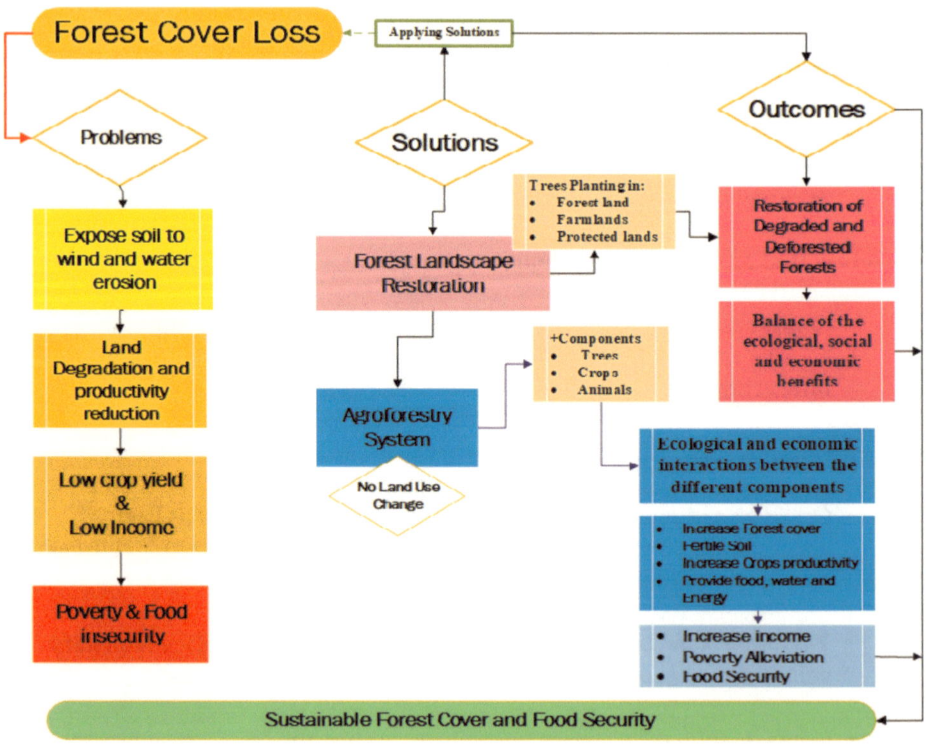

Figure 5.
Framework for Sudan forest landscape restoration [17].

well as potential restoration responses and their outcomes. The framework begins by identifying the root causes of degradation, which can be both natural and anthropogenic, and traces these to their immediate and long-term ecological and socio-economic consequences. It then outlines various restoration responses that could be applied, categorizing them into indirect approaches, such as policy changes or community engagement, and direct interventions like reforestation or soil stabilization. Finally, the framework emphasizes the necessity for robust monitoring and evaluation strategies to assess the effectiveness of these restoration efforts. This multidimensional approach aims to guide practitioners, policymakers, and researchers in the quest to address land degradation in a holistic manner.

Figure 5 outlines an Agroforestry System Framework tailored for the restoration of forest landscapes in Sudan. This comprehensive framework integrates agriculture, forestry, and local community needs to create a sustainable land management strategy. It begins with an assessment phase that gauges the current conditions of both agriculture and forestry sectors, identifying key issues such as soil degradation, deforestation, and community livelihood needs.

8. Conclusion

The urgent degradation of forest resources in Sudan's arid landscapes presents a complex, multi-layered crisis demanding swift and comprehensive action. This detailed review illustrates that the drivers of this degradation are varied and far-reaching, incorporating natural influences as well as human activities and policy-related decisions. The fallout of such degradation has ramifications beyond forest borders, impacting crucial areas like food security, biodiversity, land and water conservation, and also contributing to overarching global issues like climate change. To tackle this pressing dilemma, a dual-faceted strategy is indispensable. On one hand, immediate proactive steps must be taken to prevent the further decline of these vulnerable dryland forests. This includes a deep dive into the root causes, which span from unsustainable land management to policy shortfalls. Equally important is advocating for sustainable alternatives that carefully balance human developmental needs with ecological conservation. On the other hand, there's an urgent necessity for targeted efforts to rehabilitate already degraded forest landscapes and offset the consequent negative impacts. To ensure the efficacy of these interventions, it's critical to have a robust monitoring and evaluation mechanism in place. This system should be based on the five globally recognized indicators for assessing forest degradation, namely, biodiversity levels, productive functions, carbon storage capacities, overall forest health, and the forests' protective functions against erosion and other environmental stresses. With diligent, data-driven assessment and adaptive management, it's possible to turn the tide on these degrading trends and set Sudan's arid forests on a path toward resilience and sustainability. Given the monumental nature of the challenges at hand, collaboration emerges as a linchpin for success. It's essential to establish partnerships across local, national, and international platforms, where stakeholders can pool their collective knowledge, resources, and expertise. By working together, we can hope to not only preserve but also rejuvenate these invaluable dryland ecosystems. In doing so, we secure their essential contributions to human livelihoods, biodiversity, and overall environmental equilibrium, both within Sudan and on a global scale.

Author details

Emad H.E. Yasin[1,2*], Ahmed A.H. Siddig[1,3], Eiman E. Deiab[4], Czimber Kornel[2], Ahmed Hasoba[2,5] and Abubakr Osman[1,2]

1 Faculty of Forestry, University of Khartoum, Khartoum North, Sudan

2 Faculty of Forestry, University of Sopron, Sopron, Hungary

3 Department of Environmental Conservation, University of Massachusetts, Amherst, MA, USA

4 Institute of the Environment, National Centre for Research, Khartoum, Sudan

5 Faculty of Forest Sciences and Technology, University of Gezira Madani, Sudan

*Address all correspondence to: emadysin513@gmail.com; emad.hassanelawadyasin@phd.uni-sopron.hu

References

[1] Siddig AA. Why is biodiversity data deficiency an ongoing conservation dilemma in Africa? Journal for Nature Conservation. 2019;**50**:125719

[2] Leal Filho W, Totin E, Franke JA, Andrew SM, Abubakar IR, Azadi H, et al. Global adaptation mapping initiative team. Understanding responses to climate-related water scarcity in Africa. Science of the Total Environment. 2022; **806**:150420

[3] Mansourian S, Berrahmouni N. Review of forest and landscape restoration in Africa 2021. In: The Food and Agriculture Organization of the United Nations. Rome, Italy: African Union Development Agency-NEPAD (AUDA-NEPAD); 2021

[4] Kellner K, Risoli C, Metz M. Terminal Evaluation of the UNEP/ FAO/GEF Project: Land Degradation Assessment in Drylands (LADA). Rome, Italy: United Nations Environment Programme; 2011

[5] Kumar R, Kumar A, Saikia P. Deforestation and forests degradation impacts on the environment. In: Environmental Degradation: Challenges and Strategies for Mitigation. Cham: Springer International Publishing; 2022. pp. 19-46

[6] ITTO. ITTO Guidelines for the Restoration, Management and Rehabilitation of Degraded and Secondary Tropical Forests. ITTO Policy Development Series No. 13. Yokohama. 2002. Available from: http:// 219.127.136.74/live/Live_Server/154/ ps13e.pdf

[7] Wan Mohd Jaafar WS, Abdul Maulud KN, Muhmad Kamarulzaman AM, Raihan A, Md Sah S, Ahmad A, et al. The influence of deforestation on land surface temperature—A case study of Perak and Kedah, Malaysia. Forests. 2020;**11**(6):670

[8] Millennium Ecosystem Assessment. Ecosystems and Human Well-Being: Desertification Synthesis. Washington, DC: World Resources Institute; 2005

[9] Mateo-Vega J. Connecting local realities with global policy processes: Participatory forest biomass monitoring and scenario-based planning in Panama [Doctoral dissertation]. Canada: McGill University;

[10] Murphy PG, Lugo AE. Ecology of tropical dry forest. Annual Review Ecological System. 2022;**17**:67-88

[11] Hartung M, Carreño-Rocabado G, Peña-Claros M, van der Sande MT. Tropical dry forest resilience to fire depends on fire frequency and climate. Frontiers in Forests and Global Change. 2021;**4**:755104

[12] Bradford M, Murphy HT. The importance of large-diameter trees in the wet tropical rainforests of Australia. PLoS One. 2019;**14**(5):e0208377. DOI: 10.1371/journal.pone.0208377

[13] Janzen PCM. Species, Condiments and Medicinal Plants in SSA, Their Taxonomy and Agricultural Significance. Wageningen: Wageningen Agricultural University; 1981 Agricultural Research Reports no 906

[14] Bradford MG, Metcalfe DJ, Ford AJ, Liddell MJ, McKeown AT. Floristics, stand structure, and above-ground biomass of a 25 ha rainforest plot in the wet tropics of Australia. Journal of Tropical Forest Science. 2014;**26**(4): 543-553

[15] Solórzano A, Brasil-Machado A, Ribeiro de Oliveira R. Land use and social-ecological legacies of Rio de Janeiro's Atlantic urban forests: From charcoal production to novel ecosystems. Royal Society Open Science. 2021;**8**(6): 201855

[16] Ahmed H. Biomass Briquetting in Sudan: A Feasibility Study. New York City, New York: United States Agency for International Development Women's Refugee Commission; 2010. p. 16

[17] Gadallah NA, Adewole NA, Ajayi DD. Potential of agroforestry as forest landscape restoration tool to solve forest cover loss cum food security in Sennar and Gedaref States, Sudan. International Journal of Development and Sustainability. 2019;**8**(2):199-210

[18] Yasin EH, Kamil OH, Mulyana B. Multi-temporal satellite images analysis for assessing and mapping deforestation in Um Hataba Forest, South Kordofan, Sudan. Journal of Sylva Indonesian. 2022;**5**(01):81-92

[19] Hemida M, Yasin EH, Kheiry MA, Hammad ZM, Vityi A. Assessment of Taungya agroforestry system in dryland forests rehabilitation in Sudan. Journal of Degraded & Mining Lands Management. 2023;**10**(3):4495-4507

[20] Janzen DH. Tropical dry forests: The most endangered major tropical ecosystem. In: Wilson EO, editor. Biodiversity. Washington: National Academic Press; 1986. pp. 130-137

[21] Pollini J. Perceptions and representations of deforestation in Madagascar: From cognitive dissonances to convergences. In: Perceptions and Representations of the Malagasy Environment across Cultures. Cham: Springer International Publishing; 2023. pp. 163-189

[22] Teketay D. Deforestation, wood famine, and environmental degradation in Ethiopia's highland ecosystems: urgent need for action. Northeast African Studies. 2001;**1**:53-76

[23] FAO. The conservation of forest to another land use or long-term reduction of tree canopy cover below the minimum 10 percent threshold, Definitional issues related to reducing emissions from deforestation. 2001

[24] Bajracharya S. Community Carbon Forestry: Remote Sensing of Forest Carbon and Forest Degradation in Nepal. M.SC Thesis. the Netherlands: International Institute for Geo-information Science and Earth Observation (ITC). University of Southampton (UK); 2008

[25] FAO. FRA 2000 – On Definitions of Forest and Forest Cover Change. FRA Working Paper 23, FAO Forestry Department. 2000. Available from: https://citeseerx.ist.psu.edu/document?repid=rep1&type=pdf&doi=a88a00b108130ec6f64d0710270d7e4060d78c7f

[26] FAO. Director's Report to GCOS Steering Committee, Steering Committee Meeting III Document 11 Version 3, 25–27 January 2008. Rome, Italy: FAO Forestry Department; 2006. Available from: www.fao.org/gtos/doc/2006-GTOS-SC/11-TCO-documents-GTOS-SC-v3.doc

[27] Grainger A. Controlling Tropical Deforestation. London: Earthscan Publications Ltd; 1993. p. 310

[28] UNEP/CBD/SBSTTA. Forest Biodiversity Definitions Indicative definitions taken from the Report of the ad hoc technical expert group on forest biological diversity. 2001. Available from: http://www.biodiv.org/programmes/areas/forest/definitions.asp

[29] ITTO. Revised ITTO Criteria and Indicators for the Sustainable Management of Tropical Forests Including Reporting Format. Yokohama, Japan: ITTO Policy Development series No 15; 2005

[30] IPCC. ibidem; Alternative 3, p. 14. IPCC. ibidem; Alternative 5, p. 15. IPCC. ibidem; "framework for a definition"; 2003; p. 16

[31] FRA. Global Forest Resources Assessment. Rome: Forestry Department, FAO; 2005 Country report 212: Sudan

[32] Lund HG. Definitions of Old Growth, Pristine, Climax, Ancient Forests, Degradation, Desertification, Forest Fragmentation, and Similar Terms. Gainesville, VA: Forest Information Services; 2007. Available from: http://home.comcast.net/~gyde/pristine.htm

[33] World Bank. A World Bank Policy Paper. The Forest Sector. 1991. Available from: http://www.wrm.org.uy/actors/WB/1991policy3.html

[34] Lambin EF. Monitoring forest degradation in tropical regions by remote sensing: Some methodological issues. Global Ecology and Biogeography. 1999;**8**:191-198

[35] People & the Planet. People and Forests Glossary; 2008

[36] FAO. Proceedings. Second Expert Meeting on Harmonizing Forest-related Definitions for Use by Various Stakeholders. September 11–13, 2002

[37] IPCC. ibidem; Alternative 3, p.14. IPCC. ibidem; Alternative 5, p.15. IPCC. ibidem; "framework for a definition"; 2003. p. 16

[38] Etongo BD. Deforestation and Forest Degradation in Southern Burkina Faso: Understanding the Drivers of Change and Options for Revegetation. Finland: University of Helsinki; 2016

[39] International Union on the Conservation of Nature (IUCN). Land Degradation Neutrality: Implications and Opportunities for Conservation, Technical Brief Second Edition 27/08/2015. Nairobi: IUCN. 2015; p. 19

[40] Chalise D, Kumar L, Spalevic V, Skataric G. Estimation of sediment yield and maximum outflow using the IntErO model in the Sarada river basin of Nepal. Water. 2019;**11**(5):952

[41] EFAP. Ethiopian Forestry Action Program (EFAP). Addis Ababa: EFAP; 1994

[42] IPBES. Summary for Policymakers of the Thematic Assessment Report on Land Degradation and Restoration of the Intergovernmental Science-Policy Platform on Biodiversity and Ecosystem Services. Bonn Germany; 2018

[43] FAO. Forest Department. Global Forest Resources Assessment, Country Report. FRA2010/163. FRA Working Paper 163. Rome, Italy: FAO Forestry Department; 2010. Available from: https://www.fao.org/3/i1757e/i1757e.pdf

[44] Dash JP, Watt MS, Pearse GD, Heaphy M, Dungey HS. Assessing very high-resolution UAV imagery for monitoring forest health during a simulated disease outbreak. ISPRS Journal of Photogrammetry and Remote Sensing. 2017;**131**:1-14

[45] Souza C Jr, Siqueira J, Sales M, Fonseca A, Ribeiro J, Numata I, et al. Ten-year Landsat classification of deforestation and Forest degradation in the Brazilian Amazon. Remote Sensing.

2013;**5**(11):5493-5513. DOI: 10.3390/rs5115493

[46] Lawley V, Lewis M, Clarke K, Ostendorf B. Site-based and remote sensing methods for monitoring indicators of vegetation condition: An Australian review. Ecological Indicators. 2016;**60**:1273-1283

[47] Lausch A, Bastian O, Klotz S, Leitão PJ, Jung A, Rocchini D, et al. Understanding and assessing vegetation health by in situ species and remote-sensing approaches. Methods in Ecology and Evolution. Aug 2018;**9**(8):1799-1809

[48] Muchoney DM, Haack BN. Change detection for monitoring forest defoliation. Photogrammetric Engineering and Remote Sensing. 1994;**60**:1243-1251

[49] Batcheler CL, Craib DG. A variable area plot method of assessment of methods. New Zealand Journal of Ecology. 1985;**8**:83-95

[50] Wang C, Qi J, Cochrane M. Assessment of tropical forest degradation with Canopy fractional cover from Landsat ETM+ and IKONOS imagery. Earth Interactions. 2005;**9**(22):1-18

[51] FAO. National Forest Monitoring and Assessment – Manual Collection. Version 2.3. National Forest Monitoring and Assessment Working Paper NFMA 37/E. Rome; 2009

[52] Schomaker ME, Zarnoch SJ, Bechtold WA, Latelle DJ, Burkman WG, Cox SM. Crown condition classification: A guide to data collection and analysis (p. 78). Asheville, NC: U.S. Department of Agriculture, Forest Service, Southern Research Station, Gen. Tech. Rep. SRS-102; 2007. p. 78

[53] Will-Wolf S. Analyzing Lichen indicator Data in the Forest Inventory and Analysis Program. Portland, OR: U.S. Department of Agriculture, Forest Service, Pacific Northwest Research Station. Gen. Tech. Rep. PNW-GTR-818; 2010. p. 61

[54] Woodall CW, Monleon VJ. Sampling Protocols, Estimation Procedures, and Analytical Guidelines for Down Woody Materials Indicator of the Forest Inventory and Analysis Program. Newtown Square, PA: U.S. department of Agriculture, Forest Service, Northern Research Station, USDA Forest Service, Gen. Tech . Rep. NRS-22; 2008. p. 68

[55] Morin, Randall S, Compiler. Forest Ecosystem Health Indicators. Newtown Square, PA: U.S. Department of Agriculture, Forest Service, Northern Research Station; 2020. p. 28

[56] Stolte K, Conkling B, Campbell S, Gillespie A. Forest Health Indicators. United States Department of Agriculture (USDA); Research Triangle Park, NC: Forest Inventory and Analysis Program; 2002. Available from: http://www.fia.fs.fed.us/library/brochures/docs/Forest_Health_Indicators.pdf

[57] Schulz BK, Bechtold WA, Zarnoch SJ. Sampling Protocols, Estimation Procedures, and Analytical Guidelines for the Vegetation Diversity and Structure Indicator. Portland, OR: U.S. Department of Agriculture, Forest Service, Pacific Northwest Research Station, Gen. Tech. Rep. PNW-GTR-781; 2009. p. 53

[58] O'Neill KP, Amacher MC, Perry CH. Soils as an Indicator of Forest Health: A Guide to the Collection, Analysis, and Interpretation of Soil Indicator Data in the Forest Inventory and Analysis Program. St. Paul, MN: U.S. Department of Agriculture, Forest Service, North

Central Research Station, Gen. Tech. Rep. NC-258; 2005. p. 53

[59] Smith GC, Smith WD, Coulston JW. Ozone Bioindicator Sampling and Estimation. Newtown Square, PA: U.S. Department of Agriculture, Forest Service, Northern Research Station, Gen. Tech. Rep. NRS-20; 2007. p. 34

[60] Pandit R, Parrotta JA, Chaudhary AK, Karlen DL, Vieira DL, Anker Y, et al. A framework to evaluate land degradation and restoration responses for improved planning and decision-making. Ecosystems and People. 2020; **2020**(1):1-8

Chapter 4

Assessment and Mapping of Forest Cover Change in Dryland, Sudan Using Remote Sensing

Emad H.E. Yasin, Czimber Kornel and Mohamed Hemida

Abstract

Forest resources in the arid and semi-arid of Sudan are experiencing significant fluctuations in tree cover and ecological functionality. This study aims to bridge this gap by utilizing multi-temporal Landsat imagery and mapping forest cover change in the Nabag Forest Reserve (NFR) in South Kordofan State, Sudan. For this assessment, two cloud-free images (TM from 2011 and OLI from 2021) were downloaded and analyzed using ArcMap 10.7 and ERDAS 2014 software. Supervised classification techniques were applied, corroborated by GPS point verification and field surveys, to quantify changes in forest cover over the decade. The results revealed that dense forest cover increased from 9% in 2011 to 38.9% in 2021, while light forest cover decreased from 34.4% in 2011 to 30.9% in 2021. Additionally, the area occupied by agriculture and barren land declined from 37.2% and 19.4% in 2011 to 18.7% and 11.5% in 2021, respectively. Rapid shifts were observed in all LULC categories during the study period. The primary causes of deforestation and forest degradation were tree felling, unsustainable grazing practices, and construction activities. These findings are crucial for guiding future forest rehabilitation and creating targeted management plans for the local communities reliant on these forests.

Keywords: remote sensing, land use and land cover (LULC), forest degradation, Landsat imagery, forest rehabilitation, Nabag forest, South Kordofan state, Sudan

1. Introduction

Forest resources in Sudan's drylands are subject to variations in tree cover and ecological functionality. Reliable and up-to-date information on these resources is essential for addressing their socio-economic and environmental roles within the national environmental policy framework [1–4]. Accurate measurements of the current extent and rate of change in forest areas are imperative for devising effective management strategies to sustain the diverse ecosystem services they offer [5–9].

Forests cover over 11 billion acres in arid regions globally and are integral to local ecosystems and traditional food supply chains [10–16]. However, countries, such as Sudan, near deserts are experiencing a decline in tree cover and associated ecosystem services. Indiscriminate clearing for urbanization and economic pursuits threatens the study area's protected forests. Escalating land prices have accelerated deforestation, transforming previously forested reserves into areas for housing construction, unauthorized residential developments, and various agricultural activities. Illegal logging and firewood harvesting persist as concerns [5, 17].

For sustainable resource management, precise and current land use and cover data are indispensable. Understanding the causes and impacts of land cover changes is crucial for identifying their adverse effects on biological diversity and human development [18–22]. Remote sensing technologies offer valuable tools for monitoring these changes [23–26]. Forest composition and distribution alterations substantially influence various biological, biochemical, and ecological processes [10–12]. Data from remote sensing are often used for natural resource analysis, including tracking shifts in land use, such as forest degradation [25–28].

Temporal comparisons of satellite imagery facilitate easily identifying landscape changes [29]. Landsat data, frequently employed to monitor land cover changes at regional and global scales, helps identify and map landscape features with a high level of detail [30]. Up-to-date resource inventories are vital for effective land use planning and sustainable management [31]. By integrating remote sensing, GIS, and landscape metrics, researchers can achieve more spatially consistent outcomes, better pinpointing the social and biophysical factors behind landscape fragmentation [10, 32–34].

This study aims to assess and map temporal changes in the forest cover of Nabag Forest Reserve using multi-temporal satellite imagery, ground-truth data, and GIS integration. Information from these sources was synthesized and analyzed using matrix analysis methods to provide a comprehensive view of forest cover dynamics.

2. Material and methods

2.1 Study area

The Nabag Forest Reserve (NFR) is situated in the northern region of South Kordofan State in Sudan. Spanning 4174.2 hectares, its geographical coordinates range from 12° 30′ 0″ N to 12° 36′ 0″ N in latitude and 29° 36′ 0″ E to 29° 58′ 0″ E in longitude (**Figure 1**). The reserve lies within a low-rainfall woodland savannah zone, experiencing an annual precipitation of 350–900 mm, primarily between May and September. The area's temperature varies between 30 and 35°C.

The soil composition in the NNFR predominantly consists of clay plains interspersed with sandy clay, locally referred to as "Gardud." The vegetative cover could be more sparse and more degraded. It features a scattering of acacia trees, with Acacia seyal, Acacia mellifera, and Acacia Senegal being the dominant species. Other tree varieties, such as Balanitis aegyptiaca, are also present but less prevalent.

This setting provides a challenging backdrop for the conservation and management of the forest reserve, highlighting the importance of rigorous, science-based approaches to sustain its ecological functionality.

2.2 Detecting land use/land cover, changes

We utilized remotely sensed data from Landsat 5 and Landsat 8 satellites to assess changes in land use and land cover. These images, featuring a spatial resolution of 30 meters, were cloud-free and sourced from the United States Geological Survey's GloVis website. Specifically, we downloaded images from path 175 and row 51 for

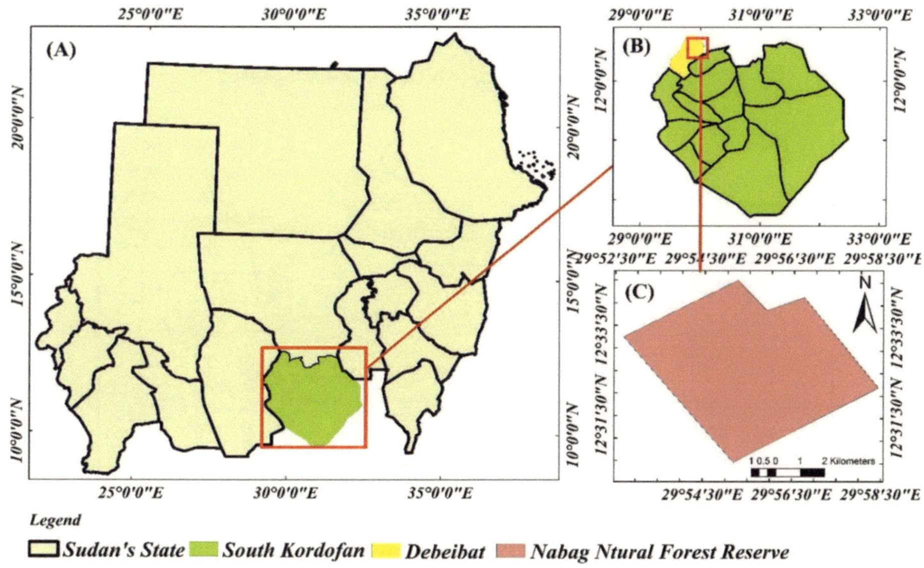

Figure 1.
Maps for the location of South Kordofan state in the southern part of Sudan (A) and the location of Nabag natural Forest reserve (study area) in South Kordofan state (B), as well as the overview of Nabag natural Forest reserve (C).

Satellite name	Sensors	Resolution	Acquisition date	Path/Row	Band used
Landsat5	TM	30 m	2011/01/11	175/051	1, 2, 3, 4
Landsat8	OLI/TRIS	30 m, 15 m	2021/01/06	175/051	2, 3, 4, 5

Table 1.
Landsat 5 TM and OLI that used in LULC determination of the study area.

2011 and 2021 (**Table 1**). Landsat 5 and 8 were selected by their geographical coverage and temporal availability, ensuring that the data would be relevant and current for our study area.

Our methodology employed an integrated approach that combined remote sensing data with field information for comprehensive data collection and analysis. This integrated process is illustrated in **Figure 2** of the paper. By synthesizing satellite imagery and on-the-ground data, we aimed to generate a robust, multidimensional perspective on land use and land cover changes within the Nabag Natural Forest Reserve.

The analysis of satellite images was performed using ERDAS Imagine 2014 and QGIS (Version 3.22.1) software, supplemented by Microsoft Excel 2019 for statistical calculations. Various stages of image processing were executed, including calibration, geometrical, atmospheric corrections, layer stacking, and composite banding. These steps converted the individual bands from each year's dataset into single-layer files. After the preprocessing, sub-scenes from the larger images were extracted (clipped) to focus on the area of interest. Supervised classification was then carried out using the maximum likelihood classifier. This method was employed to classify land use and land cover (LU/LC) in the acquired Landsat images from 2011 and 2021. To evaluate the accuracy of the classification, we relied on ground truth data, Google Earth imagery, and preexisting knowledge of the study area. Quantitative metrics,

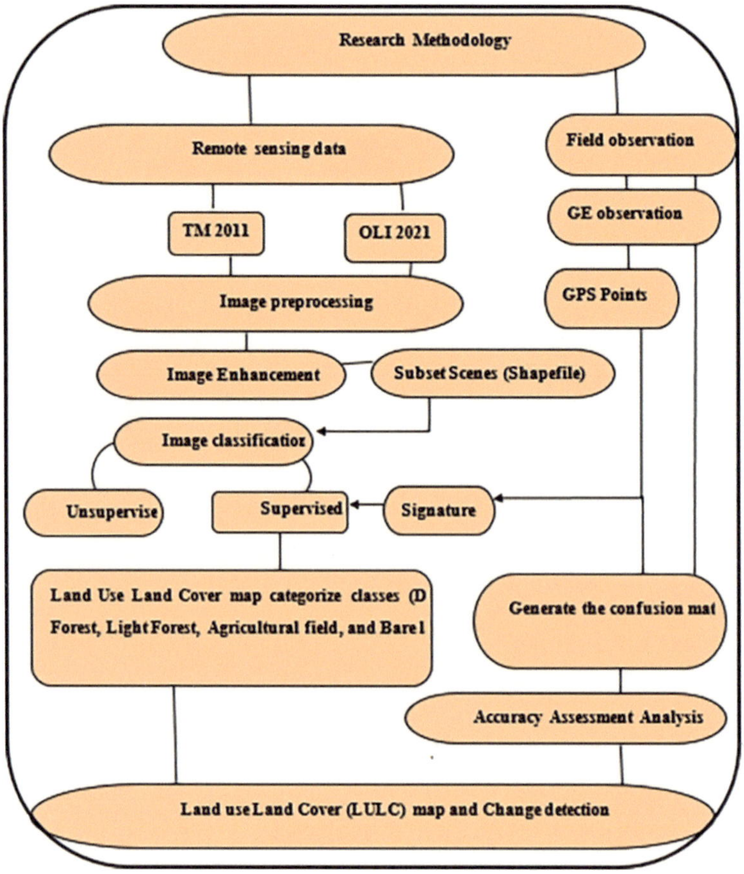

Figure 2.
Flowchart for data collection and analysis.

including users' and producers' accuracies, overall accuracy, and Kappa coefficients, were calculated to assess the reliability of the classification. To evaluate the accuracy of the classification, we relied on ground truth data, Google Earth imagery, and pre-existing knowledge of the study area. Quantitative metrics, including users' and producers' accuracies, overall accuracy, and Kappa coefficients, were calculated to assess the reliability of the classification. The results demonstrated that each classified image's overall accuracy and Kappa coefficients exceeded 85%. This high level of accuracy underscores the classification process's reliability, confirming the approach's utility in capturing the dynamics of land use and land cover changes within the Nabag Natural Forest Reserve (**Tables 2** and **3**).

3. Results and discussion

3.1 Land use/land cover results in 2021

The supervised image classification identified four distinct land cover classes: Dense forest, light forest, agriculture, and bare land. These categories are visually represented in **Figure 3**. According to the OLI 2021 classification results, dense forests

Class	Bareland	Agriculture	Light forest	Dense forest	Total	UA %
Bare land	47	7	1	3	58	81
Agriculture	4	119	3	3	129	92.3
Light forest	1	1	26	1	29	89.7
Dense forest	0	0	5	29	34	85.3
Total	52	127	35	36	250	
PA % OA = 88.4% Cappa = 82%	90.4	93.7	74.3	80.6		

Table 2.
Accuracy assessment of classified map of 2021.

Class	Bare land	Agriculture	Light forest	Dense forest	Total	UA %
Bare land	85	13	2	0	100	81
Agriculture	5	89	6	0	100	83
Light forest	0	0	29	6	35	82.9
Dense forest	0	0	0	15	15	100
Total	90	102	37	21	250	
PA % OA = 80% Cappa = 81%	94.4	87.3	78.4	71.4		

Table 3.
Accuracy assessment of classified map of 2011.

accounted for 38.9% of the total area, equivalent to 1623.78 hectares. Light forest occupied 30.9%, translating to 1289.97 hectares. Agricultural land constituted 18.7% of the area, or 780.75 hectares, while bare land comprised 11.5%, totaling 479.97 hectares. These figures and their corresponding percentages are further detailed in **Table 4**.

This classification provides a comprehensive up-to-date understanding of land use and land cover within the Nabag Forest Reserve. The results offer valuable insights into the current state of the forest and the broader landscape, essential for informed decision-making in conservation and land management strategies.

3.2 Assessment of land use/land cover change

Globally, changes in land use and land cover (LULC) are among the most significant and enduring modifications to the Earth's surface [35, 36]. Monitoring these changes is essential for generating baseline thematic maps and facilitating ongoing assessments. In the current study, as illustrated in **Figure 4** and **Table 5**, substantial LULC changes were observed during the study periods of 2011 and 2021, attributable to both human and environmental factors.

Specifically, the results reveal a significant increase in dense forest cover, growing from 9% in 2011 to 38.9% in 2021—an impressive net change of 29.9% and an annual growth rate of 2.99%. Studies on forest rehabilitation indicate that tree planting typically accelerates the recovery of forest vegetation [37]. Hemida et al. [38] found

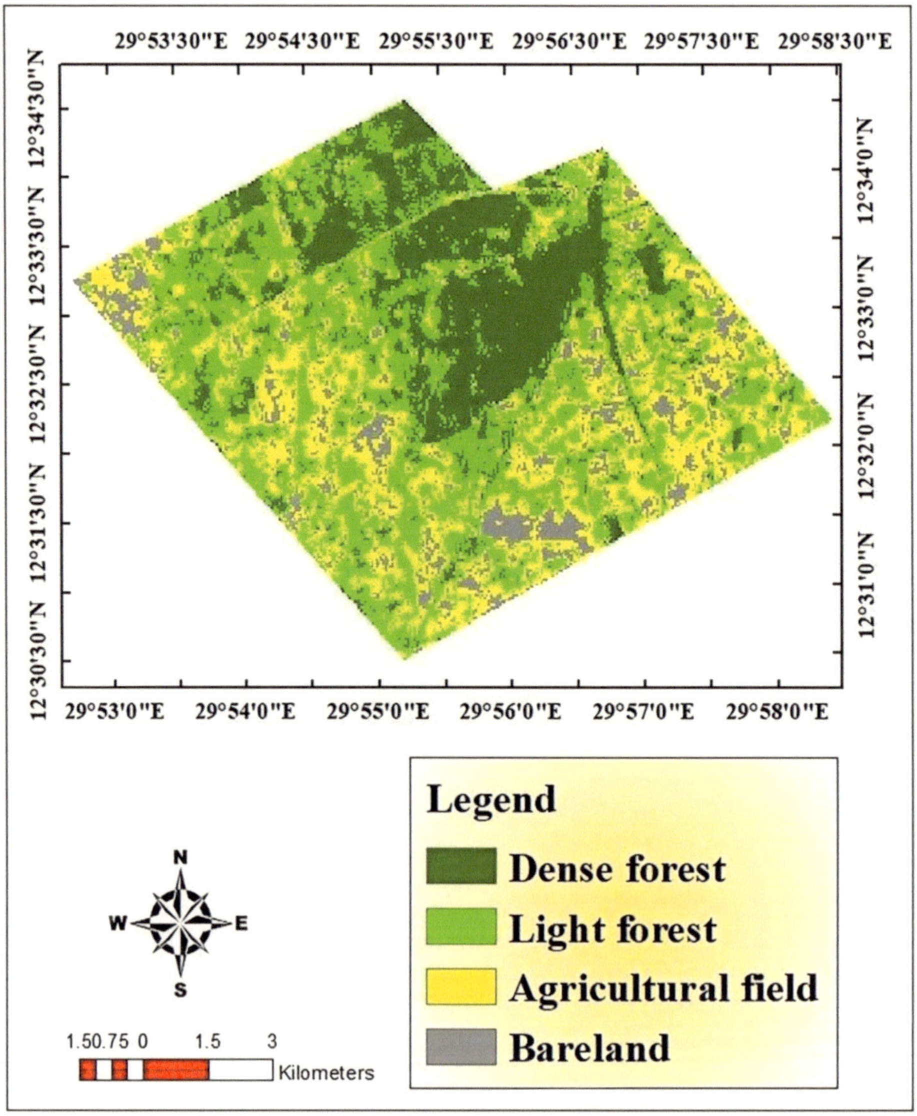

Figure 3.
Distribution of LU/LC classes detection in NFR in 2021.

Class Name	2021	
	Area (ha)	%
Dense forest	1623.78	38.9
Light forest	1289.97	30.9
Agriculture	780.75	18.7
Bare land	479.97	11.5
Total	4174.11	100

Table 4.
Distribution of LULC in the study area (ha) in 2021.

that the increase in forest cover during a certain period can be attributed to the Taungya agroforestry program initiated by the FNC in early 2005 and continuing to the present day. Salih [39] also reported that the Taungya program positively contributed to rehabilitating 3024 ha of NFR between 2005 and 2013. Similarly, a study from Nigeria highlighted the advantages of the Taungya system in promoting forest conservation and regeneration [40]. Conversely, light forest cover decreased from 34.4% in 2011 to 30.9% in 2021, with a net change of 3.5% and an annual decline rate of 0.35%. Agricultural land also reduced from 37.2% in 2011 to 18.7% in 2021, marking a net change of 18.5% and an annual decrease of 1.85%. Lastly, bare land areas contracted from 19.4% in 2011 to 11.5% in 2021, experiencing a net change of 7.9% and an annual reduction rate of 0.79%. It is important to note that other factors, including environmental, political, and agricultural influences, might have contributed to the observed changes in the reserve during this period. These factors significantly impact land use and land cover transformations [1, 41]. Historical reviews of the study region reveal that since 1982, there have been recurrent dry years. Additionally, frequent civil wars have resulted in significant losses of vegetation cover [42].

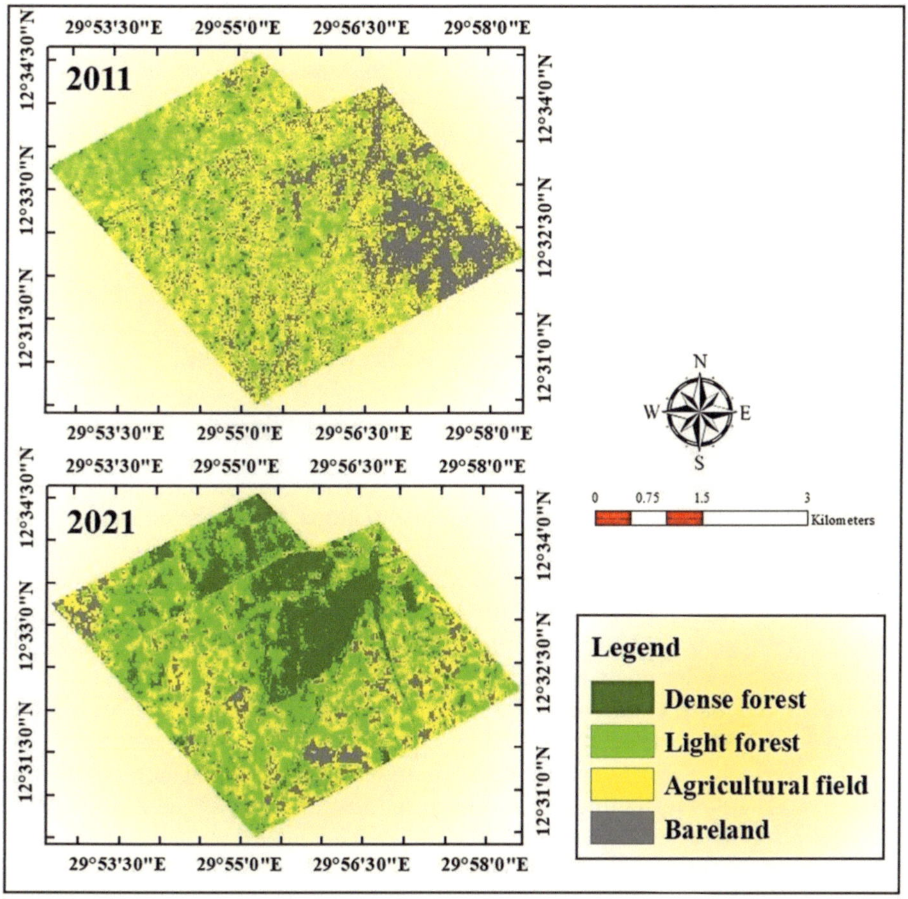

Figure 4.
Maps of land cover change in NFR between 2011 and 2021.

Class Name	2011		2021	
	Area (ha)	%	Area (ha)	%
Dense forest	377.82	9	1623.78	38.9
Light forest	1435.86	34.4	1289.97	30.9
Agriculture	1550.79	37.4	780.75	18.7
Bare land	809.64	19.4	479.97	11.5
Total	4174.11	100	4174.11	100

Table 5.
LU/LC classes and their area in percentage for 2011 and 2021.

A study by [1, 2, 38, 43] corroborates these findings, suggesting deforestation, flooding, soil erosion, and unplanned urban and agricultural expansion could lead to such LULC changes. The study also emphasizes that LULC modifications, influenced by varying environmental, political, demographic, and socio-economic conditions, are dynamic and directly affect communities living near forests.

In light of our findings, it is evident that anthropogenic activities are the driving forces behind these significant changes and disruptions, impacting the natural vegetation cycles in the study area. This research provides valuable insights into the human and environmental dimensions of LULC changes, underscoring the urgency for informed and adaptive land management strategies.

Land use and land cover (LULC) change is not merely a straightforward or linear conversion between different land types but involves a complex interplay of human activities, environmental conditions, and policy frameworks [43, 44]. This multifaceted nature of LULC makes its study particularly compelling as it can offer insights into broader sociocultural and environmental trends and processes.

Utilizing temporal remote sensing data enables the analysis of dynamic processes, emphasizing not only static snapshots of land cover at specific times but also the sequences and rates of their changes [2, 45]. In the case of the Nabag Forest Reserve, the dynamic nature of land transitions is evident spatially and temporally, especially in recent years. Use of transition matrices (as shown in **Table 4** and the maps in **Figure 5**) underscores the complex and dynamic landscape of the forest's changes. Transition matrices provide valuable analytical tools for understanding how these complex systems evolve, particularly in semiarid regions, where the conservation of natural resources is a significant concern [46].

Transition matrices allow for a multidimensional view of change over time by showing what changes have occurred and the extent and direction of these changes. In essence, they can provide insights into what drives the change in each class and help predict future trends based on past and current activities [47].

By employing matrix analysis tools, you have captured the systematic transitions in land use and land cover over the study period of 2011–2021. These matrices, visualized in **Figure 5** and **Table 6**, offer a nuanced understanding of the factors affecting Nabag forest, thereby providing a basis for more informed and targeted conservation and land management strategies.

Understanding the multilayered factors driving LULC changes is essential for developing robust and sustainable management plans. It enables stakeholders, from local communities to policymakers, to make evidence-based decisions considering the complex realities of land transformation and its implications.

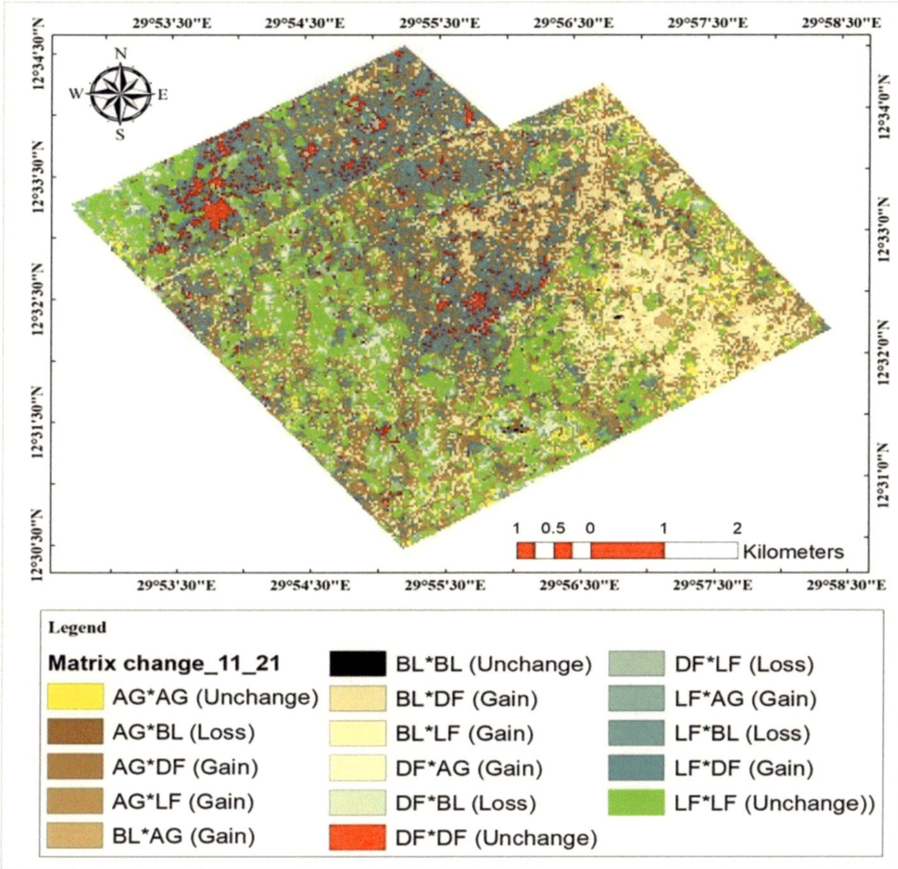

Figure 5.
LU/LC change trajectory matrix 2011–2021.

			Year	2021		
	Class name	Dense forest	Light forest	Agriculture	Bare land	Total (ha)
Year	Dense forest	179.37	668.43	570.87	205.02	1623.69
2011	Light forest	80.19	392.85	510.66	306.27	1289.97
	Agriculture	91.98	246.6	283.95	157.95	780.48
	Bare land	26.28	127.98	185.31	140.4	479.97
	Total (ha)	377.82	1435.86	1550.79	809.64	4174.11

Note: Highlighted values illustrate areas and percentages of the unchanged classes.

Table 6.
Land use and land cover trajectory matrix 2011–2021.

The intricacies of land use and land cover changes in the Nabag Natural Forest Reserve from 2011 to 2021 are highly revealing. The transition matrices offer a nuanced view, showing that the area under dense forest has seen a significant net gain of 1326.04 hectares over the decade, largely attributed to the Forest National

Corporation's agroforestry activities. However, this positive trend is counterbalanced by a loss of 118.26 hectares of dense forest, mostly converted to agriculture and bare land. This indicates that issues, such as rainfed agriculture still significantly threaten forest conservation. Furthermore, the area under light forest, categorized as fragile, has seen degradation, largely through its conversion to dense forest. This could signal environmental stress due to natural and anthropogenic factors. The decline in agricultural land could be seen as beneficial for natural ecosystems. However, it also raises concerns about declining soil fertility and the socio-economic implications for local communities. A similar decline in bare land, partially converted to dense forest, appears to be a positive trend but necessitates further inquiry into the driving factors for long-term planning. The data also flags a concerning lack of participation from local communities in forest management, emphasizing the need for more community-based, sustainable approaches. Ultimately, the transition matrix is an invaluable ledger for policymakers, offering a detailed account of losses and gains in land cover categories. Given the dynamic nature of these changes, adaptive evidence-based management strategies are crucial for the long-term conservation and sustainable development of the Nabag Forest Reserve (**Table 2**).

Sudan forests grapple with deforestation, primarily driven by agricultural growth and infrastructure projects. Despite initiatives promoting reforestation and sustainable land use, there remains a pressing need for a unified national strategy. Such a strategy should not only focus on forest conservation but also intertwine with community development, emphasizing the forest's role in climate change mitigation and biodiversity preservation (**Table 3**) [48–51].

4. Conclusion and recommendations

A multifaceted approach involving satellite imagery, geospatial technology, and ground inventories was utilized to detect, assess, monitor, and map land use and land cover (LULC) changes in the Nabag Natural Forest Reserve in South Kordofan State, Sudan. This study elucidated spatial and temporal patterns of LULC variations, offering insights for effective forest management and future rehabilitation plans. The research found substantial LULC alterations between 2011 and 2021, with differing trends and percentages. Specifically, the analysis revealed that dense forest cover increased from 9% in 2011 to 38.9% in 2021, while light forest cover declined from 34.4% to 30.9% in the same period. Agricultural and bare land areas also decreased, dropping from 37.2 and 19.4% in 2011 to 18.7 and 11.5% in 2021, respectively. These shifts were primarily driven by anthropogenic activities such as illegal tree felling, woodcutting, overgrazing, and infrastructural development. These factors contribute to both deforestation and alterations in forest cover. Such changes are anticipated to affect the forest's ecosystem services significantly. If the current degradation rate persists, the long-term sustainability of the forest could be jeopardized. Therefore, the Forestry National Corporation (FNC) should focus on preserving the remaining areas of Nabag forest. The active participation of local communities in rehabilitation efforts while respecting their interests is crucial for the success of any restoration program. Ignoring this could lead to the failure of rehabilitation initiatives. The outcomes of this study can inform the development of comprehensive forest management and rehabilitation plans that are conscious of community needs.

Acknowledgements

This chapter has been supported by the BorderEye project (TKP2021-NVA-13), which has been implemented with the support provided by the Ministry of Culture and Innovation of Hungary from the National Research, Development and Innovation Fund, financed under the TKP2021-NVA funding scheme. We also wish to express our gratitude to our colleagues for their contributions, both direct and indirect, to the data collection and analysis. The authors affirm that there are no conflicts of interest that could have affected the publication of this paper.

Author details

Emad H.E. Yasin[1,2*], Czimber Kornel[1] and Mohamed Hemida[1,2]

1 Faculty of Forestry, Institute of Geomatics and Civil Engineering, University of Sopron, Sopron, Hungary

2 Faculty of Forestry, Department of Forest Management, University of Khartoum, Khartoum North, Sudan

*Address all correspondence to: emad.hassanelawadyasin@phd.uni-sopeon.hu; emad.yasin823@gmail.com

References

[1] Gadallah NAH, Adewole NA, Ajayi DD. Potential of agroforestry as forest landscape restoration tool to solve forest cover loss cum food security in Sennar and Gedaref states, Sudan. International Journal of Sustainable Development. 2019;**8**(2):199-210

[2] Yasin EHE, Kamil OH, Mulyana B. Multi-temporal satellite images analysis for assessing and mapping deforestation in um Hataba Forest, South Kordofan, Sudan. Journal of Sylva Indonesiana. 2022;**5**(1):81-92

[3] Guirado E, Delgado-Baquerizo M, Martínez-Valderrama J, Tabik S, Alcaraz-Segura D, Maestre FT. Climate legacies drive the distribution and future restoration potential of dryland forests. Nature Plants. 2022;**8**(8):879-886

[4] Semplici G, Campbell T. The revival of the drylands: re-learning resilience to climate change from pastoral livelihoods in East Africa. Climate and Development. 2023;**3**:1-4

[5] Coulston JW, Reams GA, Wear DN, Brewer CK. Analyses of forest land use, forest land cover and changes at policy-relevant scales. Forestry. 2014;**87**:267-276

[6] Felipe-Lucia MR, de Frutos A, Comín FA. Modelling landscape management scenarios for equitable and sustainable futures in rural areas based on ecosystem services. Ecosystems People. 2022;**18**(1):76-94

[7] Lorilla RS, Kefalas G, Christou AK, Poirazidis K, Eliades NG. Enhancing the conservation status and resilience of a narrowly distributed forest: A challenge to effectively support ecosystem services in practice. Journal for Nature Conservation. 2023;**1**(73):126414

[8] Roshani SH, Kumar P, Masroor M, Rahaman MH, Rehman S, Ahmed R, et al. Forest vulnerability to climate change: A review for future research framework. Forests. 2022;**13**(6):917

[9] Nocentini S, Travaglini D, Muys B. Managing Mediterranean forests for multiple ecosystem services: Research progress and knowledge gaps. Current Forestry Reports. 2022;**8**(2):229-256

[10] FAO. Trees, Forests and Land Use in Drylands—the First Global Assessment. Rome: FAO; 2016

[11] Khan K, Hussain A, Jamil MA, Duan W, Chen L, Khan A. Alteration in forest soil biogeochemistry through coarse wood debris in Northeast China. Forests. 2022;**13**(11):1861

[12] Castañeda-Gómez L, Lajtha K, Bowden R, Mohammed Jauhar FN, Jia J, Feng X, et al. Soil organic matter molecular composition with long-term detrital alterations is controlled by site-specific forest properties. Global Change Biology. 2023;**29**(1):243-259

[13] Milton AD. Forest Resilience for Livelihoods and Ecosystem Services. USA: George Mason University; 2015

[14] Pretzsch J. Forest related rural livelihood strategies in national and global development. For Trees Livelihoods. 2005;**15**:115-127

[15] Kalogiannidis S, Kalfas D, Loizou E, Chatzitheodoridis F. Forestry bioeconomy contribution on socioeconomic development: Evidence from Greece. Land. 2022;**11**(12):2139

[16] Shebitz DJ, Agnew LP, Kerns S, Oviedo A, Ha J. Let it grow (Back): A call

for the conservation of secondary forests as medicinal plant habitat. Ethnobiology Letters. 2023;**14**(2):37-46

[17] Mohamed NAH, Bannari A, Fadul HM, Zakieldeen S. Ecological zones degradation analysis in Central Sudan during a half century using remote sensing and GIS. Advanced Remote Sensing. 2016;**5**:355-371

[18] Daily C, Ellison K. The Economy of Nature: The Quest to Make Conservation Profitable. Washington, DC: Island Press; 2002

[19] Mora-Arroyo JP, Sannchez-Azofeifa GA, Rivard B, Calvo JC, Janzen DH. Dynamics in landscape structure and composition for the Chorotega region, Costa Rica from 1960 to 2000. Agriculture, Ecosystems & Environment. 2005;**106**:27-29

[20] Msofe NK, Sheng L, Lyimo J. Land use change trends and their driving forces in the Kilombero valley floodplain, Southeastern Tanzania. Sustainability. 2019;**11**(2):505

[21] Assede ES, Orou H, Biaou SS, Geldenhuys CJ, Ahononga FC, Chirwa PW. Understanding drivers of land use and land cover change in Africa: A review. Current Landscape Ecology Reports. 2023;**3**:1-1

[22] Debnath J, Sahariah D, Lahon D, Nath N, Chand K, Meraj G, et al. Geospatial modeling to assess the past and future land use-land cover changes in the Brahmaputra Valley, NE India, for sustainable land resource management. Environmental Science and Pollution Research. 2022;**23**:1-24

[23] Singh A. Digital change detection techniques using remotely sensed data. International Journal of Remote Sensing. 1989;**10**(6):989-1003

[24] Mouat DA, Mahin GG, Landcaster J. Remote sensing techniques in the analysis of change detection. Geocarto International. 1993;**2**:39-49

[25] Musaoglu N, Saroglu E, Bektas F, Goksel C, Kaya S. Assessment of Forest Degradation by Means of Remotely Sensed Data. Proceedings of the Global Developments in Environmental Earth Observation from Space. Rotterdam: Millpress; 2006. p. 413

[26] Savastru DM, Zoran MA, Savastru RS. Satellite remote sensing detection of forest vegetation land cover changes and their potential drivers. Proceedings of SPIE. 2019;**11149**:1114926

[27] Hassan Z, Shabbir R, Ahmad SS, Malik AH, Aziz N, Butt A, et al. Dynamics of land use and land cover change (LULCC) using geospatial techniques: A case study of Islamabad Pakistan. Springerplus. 2016;**5**(1):812

[28] Tolessa T, Senbeta F, Kidane M. The impact of land use/land cover change on ecosystem services in the central highlands of Ethiopia. Ecosystem Services. 2017;**23**:47-54

[29] Abd El-Kawy OR, Rød JK, Ismail HA, Suliman AS. Land use and land cover change detection in the western Nile delta of Egypt using remote sensing data. Applied Geography. 2011;**31**(2):483-494

[30] Pax-Lenney M, Woodcock CE, Macomber SA, Gopal S, Song C. Forest mapping with a generalized classifier and Landsat TM data. Remote Sensing of Environment. 2001;**77**:241-250

[31] Ochoa-Gaona S, Gonzalez-Espinosa M. Land use and deforestation in the highlands of Chiapas, Mexico. Applied Geography. 2000;**20**:17-42

[32] Tariq A, Mumtaz F. A series of spatio-temporal analyses and predicting modeling of land use and land cover changes using an integrated Markov chain and cellular automata models. Environmental Science and Pollution Research. 2023;**30**(16):47470-47484

[33] Deb S, Imdad K, Patel PP, Sahul W, Parween S, Rashid R, et al. Approaches and methodologies on mapping vegetation cover and biodiversity status using remote sensing and spatial analysis: A systematic review. Conservation, Management and Monitoring of Forest Resources of India. 2022;**5**:379-408

[34] Parida BR, Kumar P, Chakraborty JS, Bag N, Kumar N. Dynamics of land use/land cover over Indian Himalayan region (IHR) using satellite data from 1985 to 2015 coupled with local perceptions. In: Handbook of Himalayan Ecosystems and Sustainability. Vol. 2. Abingdon, Oxon: CRC Press, Taylor & Francis Group; 2022. p. 1

[35] NDDC/UNDP. Sudan National Action Programme (Snap). Khartoum, Sudan: Republic of Sudan Ministry of Agriculture and Forestry National Drought and Desertification Control Unit (Nddcu); 2006

[36] Lewis J, Liljedahl B. Groundwater surveys in developing regions. Air, Soil and Water Research. 2010;**10**:95-104

[37] Holl KD. Restoring tropical forest. Nature Education Knowledge. 2013;**4**(4):4

[38] Hemida M, Yasin EHE, Kheiry MA, Hammad ZM, Vityi A. Assessment of Taungya agroforestry system in dryland forests rehabilitation in Sudan. Journal of Degraded & Mining Lands Management. 2023;**10**(3):4495-4507

[39] Salih BA. Rehabilitation of Nabag Forest through Taungya System - ElQoz Locality Southern Kordofan State [Thesis]. Sudan: Sudan University of Science and Technology; 2013. p. 39

[40] Azeez IO, Popoola L, Shulamite IO. An assessment of the forest regeneration potential of the taungya system of farming in Oyo state, South-Western Nigeria. Southern Forests: Journal of Forest Science. 2017;**79**:143-151. DOI: 10.2989/20702620.2016.1255425

[41] Hosonuma N, Herold M, De Sy V, De Fries RS, Brockhaus M, Verchot L, et al. An assessment of deforestation and forest degradation drivers in developing countries. Environmental Research Letters. 2012;**7**(4):044009. DOI: 10.1088/1748-9326/7/4/044009

[42] Deafalla THH, Csaplovics E, El-Abbas MM. Environmental change impacts and human security in semi-arid region, the case of Nuba Mountains of Sudan. In: Behnassi M, Gupta H, Pollmann O, editors. Human and Environmental Security in the Era of Global Risks. Cham: Springer; 2019. DOI: 10.1007/978-3-319-92828-9_10

[43] Arfat Y. Land Use/Land Cover Change Detection and Quantification: A Case Study in Eastern Sudan. Lund: Lund University; 2012

[44] Lambin EF, Geist HJ, Lepers E. Dynamics of land-use and land-cover change in tropical regions. Annual Review of Environment and Resources. 2003;**28**(1):205-241

[45] Briones PS, Sepúlveda-Varas A. Systematic transitions in land use and land cover in a pre-Andean subwatershed with high human intervention in the Araucania region, Chile. Cienc Invest.

2016;**43**(3):396-407. DOI: 10.4067/S0718-16202016000300006

[46] Petit C, Scudder T, Lambin E. Quantifying processes of land-cover change by remote sensing: Resettlement and rapid land-cover changes in southeastern Zambia. International Journal of Remote Sensing. 2001;**22**(17):3435-3456

[47] Zewdie W, Csaplovics E. Identifying categorical land use transition and land degradation in northwestern drylands of Ethiopia. Remote Sensing. 2016;**8**(5):408. DOI: 10.3390/rs8050408

[48] El Tahir BA, Fadl KEM, Fadlalmula AGD. Forest biodiversity in Kordofan region, Sudan: Effects of climate change, pests, disease and human activity. Biodiversity. 2010;**11**:34-44. DOI: 10.1080/14888386.2010.9712662

[49] Mamoun HO, El Zein A. Species dynamics and potential disturbances in El Nour Natural Forest reserve, Sudan. Journal of Forest Products and Industries. 2012;**1**(2):10-20

[50] El Gunaid F, Hassan AMH, Dafalla E. Effect of human activities on forest biodiversity in white Nile state, Sudan. International Journal of Innovation and Applied Studies. 2013;**2**(4):547-555

[51] Gadallah NAH, Taha ISA, Hano AIA, Siddig AAH, Bo HJ. Integrated approach for assessment and monitoring of forests conditions in the drylands of Sudan. Arid Ecosystems. 2022;**12**:142-153. DOI: 10.1134/S2079096122020032

Section 3

Restoration of Mountain Ecosystems under Global Climate Change

Chapter 5

Perspective Chapter: Ways to Improve the Ecological State of Abandoned Quarries in Countries with Mountainous Terrain

John Asatryan

Abstract

The wall stone of a correct form is one of the main construction materials used in Armenia and other countries with mountainous terrain. The current technology of stone cutting using rail-mounted machines is ineffective due to various reasons. These include excessive cutting forces, vibrations, formation of chips and cracks, incomplete cutting of benches resulting in remnants left at the quarry bottoms, and significant accumulation of waste. These problems have severe environmental consequences. In countries with mountainous regions and limited quarry lengths, the problem of inefficient stone cutting is even more pronounced. Over the past few decades, and particularly after the devastating earthquake of 1988, stone quarries have been operated without considering environmental regulations. Many quarries have been abandoned as well. The author proposes two approaches to improve the mountain ecosystem: the conversion of abandoned quarries into a connected network of small reservoirs and the development of new trackless stone-mining machines equipped with wheels instead of rails. These innovations enable the extraction of the wall stones minimizing waste and reducing harm to the ecosystem. The addition of forest zones around the reservoirs can further mitigate the environmental impact of quarries by reducing stone dust release into the atmosphere.

Keywords: stone quarry, abandoned quarry, stone quarrying machine, wheeled machine, circular saw

1. Introduction

The wall stone of a correct form is one of the main construction materials used in Armenia and other countries with mountainous terrain.

For decades in the Countries of the Independent States (CIS) of the former Soviet Union and neighboring countries, the extraction of wall stone of the correct form with cross dimensions of 200 × 300 mm was carried out by rail machines CMP-026/1.

1.1 Problems

The efficiency of these machines is quite low because their design, tooling and technology of stone extraction have exhausted all their technical possibilities for improvement. High cutting forces during the extraction of rocks with a strength of up to 40 MPa, lead to vibrations in the multi-ton machines, causing formation of chips and cracks and large mining waste causing significant environmental damage. Quarry saws emit clouds of stone dust into the air, especially in quarries located in windy areas, posing severe natural hazards for nearby settlements. The installation of rails at different heights creates underused stepped sections within the quarry. The saw blades of the machine have insufficient durability due to the increased strength of the lower layers of the quarry.

These issues are particularly challenging for countries with mountainous landscapes, where quarry lengths are relatively small. Therefore, improving the efficiency of stone extraction becomes a complex and demanding task. **Figure 1** provides a typical illustration of the stone extraction process using a CMP-026/1 rail machine.

Figure 2 depicts an abandoned quarry, which highlights the issue of accumulated waste mountains resulting from stone extraction. These waste piles, along with the abandoned quarries, pose a significant environmental threat and the risk of a potential catastrophe for future generations. Therefore, it is crucial to address and resolve the challenges related to the protection, utilization, and restoration of mountain ecosystems. Immediate solutions are needed to tackle these pressing issues.

Given the small land area of the Republic of Armenia and other comparable countries, it is essential to formulate a well-defined strategy for the development of the stone industry. This strategy should be built upon the utilization of modern scientific and engineering advancements that prioritize minimizing environmental harm.

Figure 1.
View of machine and quarry when extracting a stone of the correct form.

Figure 2.
Landscape view of an abandoned quarry in a country with mountainous terrain.

2. Solution

The author proposes two approaches to address the conservation, exploitation, and restoration of ecosystems in countries with mountainous terrain:

1. The conversion of abandoned quarries into a connected network of small reservoirs, supplied by a central dam located at a higher geographical level than the quarries. This project can be realistically implemented over an area of 3–4 thousand square kilometers in mountainous countries. The implementation of a such project requires thorough study of the physical properties of the quarry rocks, utilization/disposal of stone mining waste both in the quarry and in its surrounding areas, as well as the establishment of forest zones covering the perimeter of the quarry.

2. The development of new trackless stone-mining machines and technologies that are equipped with wheels instead of rails. These innovative machines would enable the extraction of the wall stones of the correct form with minimal waste and without causing harm to the ecosystem. Furthermore, they would allow for the utilization of abandoned quarries by extracting the remaining hard and unexploited layers of rock located at lower levels.

2.1 Transformation of abandoned quarries

Research and development efforts have been undertaken in the first approach, focusing on the Shirak region of Armenia. The following steps have been taken:

a. Identification of a feeding dam with the highest geographical location to serve as the water source for a network of small reservoirs-quarries. In this case, the newly constructed Caps reservoir is considered. It has a water volume of 61 million cubic meters and is situated at an altitude of 1730 meters above sea level. The dam is being built on the Akhuryan River, which originates from Lake Arpi at an elevation of 2017 meters and flows into the Araks River at an elevation of 905 meters.

b. Compilation and collection of maps of both active and abandoned quarries in the region. The selection of quarries to be included in the network of small reservoirs is based on several parameters, including their location, altitude, contour, dimensions, depth, type of stone, composition, and physical and mechanical properties.

c. Preliminary assessment of waste utilization possibilities for each selected quarry and its surroundings. Suitable crushing, screening, and dispersion techniques are employed to process the waste into crushed stone, sand, and powders. The potential uses of stone powders are determined, such as their application as lightweight and active fillers in concrete, fillers in the production of plastic and rubber, heat-insulating materials, acid-resistant products, soil fertilizers due to their filtration properties, and for polishing materials with dimensions of 0.05 mm or smaller.

d. A network of water supply channels to individual reservoirs-quarries is being developed in advance.

The map of the Shirak region shows the places of stone quarries (**Figure 3**).

There are two designated zones. The first zone covers the quarries located below the level of the Caps reservoir. Total area of those quarries, along with the surrounding forest zones covers about a third of the area of the region.

Out of the quarries assessed, approximately 50% are currently in operation. A network of small reservoirs has been established, incorporating 17 abandoned quarries. The central water supply channel, represented by a dotted line, is still undergoing refinements as the project continues to develop.

Figure 3.
Map of stone quarries in the Shirak region: Δ – quarries included in the zone of the network of reservoirs; Δ – quarries of high mountain zones.

The second area focuses on the quarries located in highland regions. Efforts to preserve, utilize, and restore highland ecosystems are partially being addressed through the pilot grant program called "Waste Management of Closed Stone Quarries and Flooding of the City of Artik." This program is initiated by the Ministry of the Environment and funded by the United Nations Adaptation Fund. As part of this initiative, forest parks and recreational areas are being established on closed quarries, and efforts are being made to restore cultivated lands.

The overall outcome of the aforementioned initiatives will involve conducting research and assessing the environmental impact of the entire network of reservoirs once the project is completed. It is anticipated that the system of small reservoirs will have a substantial positive influence on improving the regional environment. The emission of stone dust into the atmosphere will notably decrease, the climate will become milder, the water basin of the Akhuryan River will improve, and there will be an expansion of recreational areas for the local population, among other benefits. These effects will be examined and evaluated as part of the post-project assessment.

2.2 Replacement of the machinery

The author sees the second approach for the cardinal improvement of the ecosystems of mountainous regions, which involves moving away from the utilization of rail-based stone mining machines in the extraction wall and facing stones. Presently, various countries, including Armenia, Russia, China, and Italy, produce and employ different rail-mounted stone mining machines. However, these machines suffer from similar drawbacks as mentioned earlier.

The solution of this problem is in the development of the new wheeled trackless stone mining machines and technologies that ensure the extraction of wall stone of the correct form with minimal waste, without harming the ecosystem. Patent solutions to these problems and preliminary results of engineering developments and tests confirm the correctness of the chosen direction [1–5].

Analysis of the dimensions of stones of the correct form showed that currently, the stones are predominantly mined with cross dimensions of H × B generally not exceeding 200(250) × 250(300) mm, with the lengths ranging from10 to 400 mm. Moreover, the purpose of the wall stone also has changed. They are now primarily utilized for decorative exterior cladding with beautiful "rock" relief rather than serving as load-bearing elements.

To extract stones with the aforementioned parameters, a special wheeled machine has been developed. This machine aims to significantly reduce cutting forces, minimize vibrations, eliminate the formation of chips and cracks, and enable operation at higher cutting speeds and feeds. These improvements are achieved through the optimization of operational parameters and instrumental characteristics of the machine.

Transitioning from the conventional rail-based machine design to a wheeled and highly maneuverable design opens up opportunities for applying new and more efficient technological processes. The development of new stone cutting machines offers significant prospects for incorporating modern hydraulic units of the latest generation. These hydraulic units can be utilized not only in the drives of the stone-cutting heads but also in the hydrostatic transmissions of the machine for feed motion.

The new machine features saw blades positioned at the edges of the machine, with their cutting surfaces covering the transverse width L_m of the machine. In this prototype machine, the distance l_p between adjacent saws is set to be twice the length L of the stone being cut, taking into account the width b_p of the cut, i.e. $l_p = 2(L + b_p)$.

This configuration is preferable since lp is a multiple of the L_m width of the machine. As a result, the machine undergoes a periodic displacement/offset by the length of the stone before each new transverse passage.

The primary data used for designing the new machine were as follows:

a. The strength of the stone being cut was up to 40 MPa.

b. Circular saws with diameters of 800 and 630 mm, and a width of 12 mm, equipped with cutters featuring hard alloy or diamond polycrystal plates, capable of achieving a cutting depth of up to 250 mm.

c. The calculated power and torque values for one saw were 10 kW and 1200 Nm, respectively.

d. The maximum cutting speed and feed were set at 6 m/s and 5 m/min, respectively.

e. Solid tires were used instead of tubes in the machine's construction.

The project was conducted collaboratively between the Gyumri branch of the National Polytechnic University of Armenia and the Open Joint Stock Company "Karatmekena". Their efforts led to the development of a completely innovative machine capable of performing both transverse and longitudinal technological operations, as well as their combinations. The hydraulic units utilized in the machines were sourced from *Poclain Hydraulics*. A prototype of the new machine, featuring combined operations, was created using the forklift model 4014 from the Lvov plant "Avtopogruzchik," which has a load capacity of 5 tons. The selection of hydraulic units was made in consultation with a representative from Poclain Hydraulics.

Figure 4 showcases the newly developed combined machine designed specifically for extracting a stone block of the correct form with cross dimensions of 200 × 300 mm. The machine follows a sequence of operations that involve:

a. Cross passages when the horizontal cutting head is raised upwards.

b. Longitudinal passages when the transverse saw blades are removed and only one is left at the left cutting head. This is the stage when the separation of stone from the massif occurs.

An important aspect of the new wheeled machine is its ability to maintain straight-line movement, despite potential deviations caused by solid inclusions in the rock or external factors. To address this, the machine is equipped with an electronic control system comprising a three-axis gyroscopic servo system, a servo amplifier, an electromagnetic hydraulic distributor, and a parallel-connected reverse portioner to the power steering mechanism. This system, along with an external device responsible for adjusting straight-line movement, continuously monitors the machine's motion and intervenes promptly in case of any deviations.

The flexibility offered by this machine in performing various technological operations presents a wide range of opportunities for repurposing abandoned quarries. Such machines will prove highly efficient for the secondary utilization of these quarries, which are numerous and have been environmentally harmful. The elimination of such residues will significantly contribute to the improvement of mountainous ecosystems.

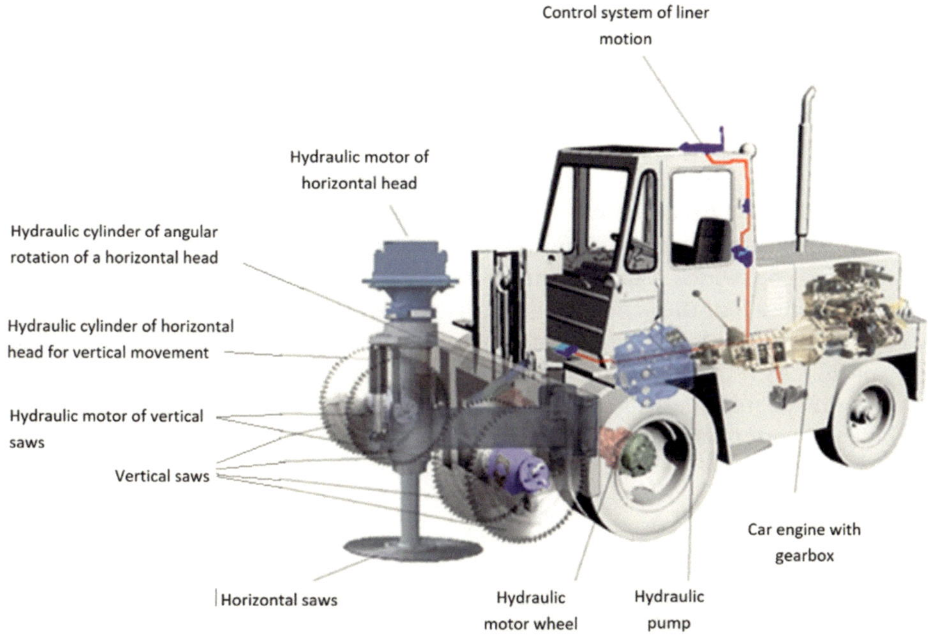

Figure 4.
Combined machine for the extraction of regular shape rock-stone.

The technology of stone extraction by the new machine addresses an important issue in quarrying: the formation of residual stepped protrusions. By utilizing edge saw blades covering the transverse width of the machine, it is able to create smooth, vertical walls without protrusions as it moves along in successive longitudinal passes.

Figure 5 depicts the cross-sectional profile of an abandoned quarry formed after the extraction of stone by the CMP-026/1 rail machine. This profile exhibits a stepped formation along the quarry's perimeter, extending down to the lower layers. The

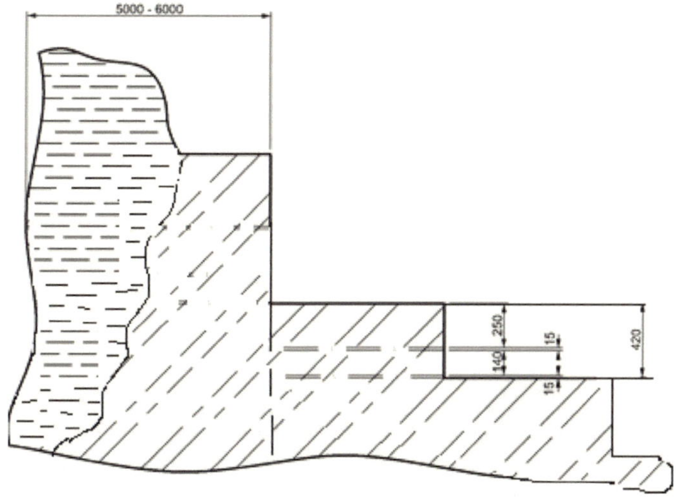

Figure 5.
Cross profile of an abandoned quarry and a usage scheme.

steps have dimensions of 420 × 1000 mm. Approximately 30–40% of the original stone volume remains in the quarry. The image showcases the profile of the quarry after mining several layers using the new machine, resulting in a vertical wall with no remaining residues. The stone extraction process follows the same technology described earlier.

The comparison between the productivity of the new machine and the CMP-026/1 machine for extracting wall stone of the correct form has been conducted. The calculation has resulted in a relationship that determines the actual productivity of the new machine, which features four vertical saws and one horizontal saw. It has been determined that the actual productivity of the new machine will range from 2.2 to 3 cubic meters per hour. On average, this is 1.2 to 1.8 times higher than the productivity of the CMP-026/1 machine. The increased productivity is achieved through the optimization of stone cutting modes, a higher percentage of usable output, enhanced machine utilization over time, and reduced time spent on auxiliary operations.

3. Conclusion

1. Creating a network of small reservoirs on abandoned stone quarries, surrounded by forest zones, is a viable approach to improving the condition of mountain ecosystems in countries with mountainous terrain.

2. The utilization of new wheel-based stone-mining machines instead of rail-based machines shows promise as it minimizes waste generation. The flexibility of maneuvering these machines in the quarry allows for the extraction of stone from untouched lower layers, enabling the secondary use of abandoned quarries.

Author details

John Asatryan
National Polytechnic University of Armenia, Gyumri, Armenia

*Address all correspondence to: eaagb@yandex.ru

References

[1] Asatryan J, Mkttchyan N, Msryan F, Arutyunyan S. Stone extraction machine. Patent RA 2841 A. 2014

[2] Asatryan J, Mkttchyan N, Msryan F, Arutyunyan S. Stone extraction machine. Patent RA 2861 A. 2014

[3] Asatryan J, Harutyunyan S. Optimal Transmissions for Wheeled Sone Mining Machines. Mechanical Engineering Solutions. Vol. 2. Yerevan: NPUA (National Polytechnic University of Armenia), AUA (American University of Armenia); 2018. pp. 20-26

[4] Asatryan J, Harutyunyan S. Machine and technology of cutting regular-shape face stone in mountainous countries. Gornyy Informatsionno-Analiticheskiy Byulleten. 2018;7:89-98

[5] Asatryan J, Harutyunyan S. Technology and a wheel machine for surplus quarrying wall stone of a correct form. In: Proceedings of the RA NAS and NPUA. Vol. 68. Yerevan; 2015

Chapter 6

Protection, Development, and Restoration of Mountain Ecosystems: Taking the Wugong Mountain Meadow Ecosystem as a Study

Zhi Li, Yanmei Wang, Xiaodong Geng, Qifei Cai, Zhen Liu, Ling Zhang and Xiaomin Guo

Abstract

In this comprehensive chapter, we focus on the case study of a typical mountain meadow ecosystem within Wugong Mountain; and provide imperative aspects of protection, development, and restoration of such ecosystems. Firstly, the importance of protecting mountain meadow ecosystems was introduced. Mountain meadow ecosystems are an important component of global life support systems and ecosystems and are crucial to the global ecosystem. Secondly, the various challenges faced by mountain meadow ecosystems, including environmental degradation and disaster risks, were discussed, and special measures were also needed to adapt and mitigate climate change. Subsequently, a review was conducted on the research and protection of mountain meadow ecosystems, including research on vegetation biodiversity and the relationship between mountain soil and plant growth. In addition, the restoration of mountain meadow ecosystems was also discussed, including determining restoration goals, conducting comprehensive investigations of soil and environmental conditions, and determining vegetation planting methods. Finally, we explored the research on environmental carrying capacity in the process of sustainable development, including limiting development and developing sustainable tourism. In summary, we have systematically summarized the protection, development, and restoration of mountain meadow ecosystems, providing reference and inspiration for relevant research.

Keywords: ecology, mountain areas, meadow, soil, plant, protection, sustainable development

1. Introduction

Mountain ecosystems are characterized by rich biodiversity and unique geographical environments, which play a crucial role in preserving water, conserving soil and water, regulating climate, and protecting biodiversity [1]. Thus, safeguarding these ecosystems is of significant ecological, social, and economic importance [2]. In recent years, researchers have conducted extensive research on these aspects and proposed many protection strategies and measures, such as establishing natural reserves, ecological restoration, ecological sand fixation, vegetation restoration, and so forth [3–5]. The development of these ecosystems must adhere to the principles of sustainable development and focus on the unity of ecological, social, and economic benefits. The development direction mainly includes ecotourism, ecological agriculture, forestry, clean energy, etc. Throughout the development process, prioritizing ecological protection, judicious resource utilization, and industrial structure optimization is crucial. Scholars have further delved into the development of these ecosystems, examining various models, strategies, and policy suggestions [6, 7]. Ecosystem restoration involves actions like ecological and vegetation restoration, and soil enhancement to return the ecosystem to its near-natural state. Restoration efforts must concentrate on reviving species diversity, ecosystem function, and stability. Researchers have conducted extensive research on the restoration of mountain ecosystems and proposed many restoration technologies and methods, such as ecological sand fixation, vegetation restoration, and soil improvement. The intertwined processes of protecting, developing, and restoring these ecosystems are long term and complex, spanning multiple disciplines such as ecology, geography, and environmental science [8, 9].

The Earth's three primary carbon reservoirs are its grasslands, forests, and oceans. Mountainous meadows are a larger form of terrestrial ecosystem with significant carbon storage capacity, they play a significant role in the global carbon cycle [10]. These meadows predominantly occupy flat terrains and are known for their rich organic matter and fertile soils. Recently, grassland research has emerged as a key area of interest to scientists. Exploring grassland mountain ecosystems represents an interdisciplinary research endeavor, encompassing fields such as ecology, botany, soil science, and meteorology [11]. In-depth exploration and comprehension of these ecosystems' characteristics and principles can furnish the scientific foundation and theoretical guidance necessary for their protection and restoration.

Wugong Mountain in southern China has been widely known in recent years due to its beautiful alpine meadow landscape, and it also has rich historical and cultural connotations [12]. Wugong Mountain's strategic location and orientation significantly influence the vegetation zoning across East China, serving as a natural demarcation for the central and northern subzones of evergreen broad-leaved forests in China's humid subtropical east [13]. The mountain's high elevation and steep terrain lead to a pronounced differentiation in the vertical zoning of both soil and vegetation, especially the widely distributed mountain meadows around the main peak, making it a rare natural museum of vegetation vertical zonality. Within Jiangxi, Wugong Mountain stands out as the only one with typical mountain meadow vegetation, unlike Lushan Mountain, Jinggang Mountain, and similar others [14]. Our colleagues have engaged in comprehensive studies here, including investigating and analyzing the occurrence conditions, nutrient characteristics, soil carbon, nitrogen, and soil microbial characteristics of mountain meadow soils at different elevations, to explore the nutrient characteristics, biological activity, and limiting factors of mountain meadow soils [15–18]. Utilizing advanced international methods for soil nutrient analysis, they have conducted experiments on mountain meadow soil, determining nutrient limiting factors and spatial variability in different degradation stages and elevation meadow soil. They have uncovered the varying traits of nutrients, structure, morphology, microorganisms, and soil fertility in grassland soils across different times and spaces, along with the underlying mechanisms [19, 20]. These efforts foster environmental sustainability in grassland ecosystems, and aid in the restoration and enhancement of degraded grassland soils. They provide preliminary experience for the evaluation system and methods of the health status of grasslands in Wugong Mountain.

Hence, we utilize Wugong Mountain as a representative case study to examine research on the conservation, development, and restoration of mountain ecosystems, in order to provide a theoretical basis for the protection and sustainable development of mountain ecosystems.

2. Overview of the research area

2.1 Geographical location and location characteristics

Positioned in the northern part of the Luoxiao Mountains, Wugong Mountain marks the divide between the Xiang and Gan River systems (114°10′ ~ 114°17′ E, 27°25′ ~ 27°35′ N), encompassing around 260 km^2, it's an average elevation exceeds 1500 m (**Figure 1**). Baihe Peak, also known as the Jinding, reaches the highest elevation of 1918.3 m [21]. The mountain experiences an annual average temperature of 14–16°C, with humidity levels ranging from 70 to 80%, and an annual rainfall of 1350–1570 mm. Wugong Mountain is predominantly formed of granite and gneiss. Wugong Mountain significantly influences the vegetation zoning across eastern China. Its steep, high-elevation terrain creates pronounced vertical zonation in climate, soil, and vegetation [22].

2.2 Natural resources

Wugong Mountain has beautiful scenery, with an elevation difference of 1600 m. Its main vein mainly runs in a southwest-northeast direction. The geological structure is mainly composed of granite core complex and peak cliff landforms, with an elevation generally exceeding 1000 m, and many mountain peaks reaching over 1500 m. It is a unique landscape of granite and mixed rock landforms, with the main types of

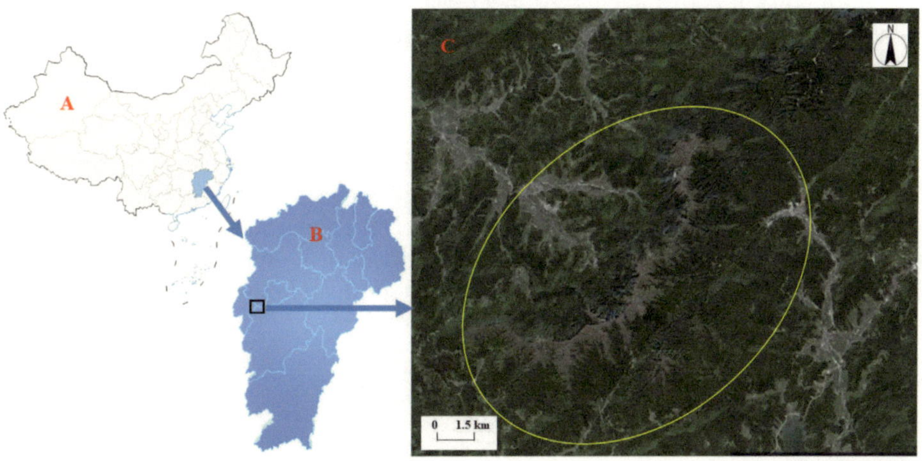

Figure 1.
The geographical location and scope of Wugong Mountain (within the yellow circle). A: Map of China, B: Map of Jiangxi Province, and C: Satellite map of Wugong Mountain area.

landforms being middle and low mountains. Red soil, yellow soil, yellow-brown soil, and plateau meadow soil are distributed in a gradient pattern along the elevation [23].

Wugong Mountain belongs to a subtropical monsoon humid climate with distinct four seasons, mild climate, abundant rainfall, and large clouds and mist. The annual average temperature is 14–16°C, and the summer is cool and pleasant. At the same time, the temperature is lower than that of Mount Lushan and Mount Huangshan, which is a good place for summer vacation. The annual precipitation of Wugong Mountain is about 1700 mm, with abundant precipitation, forming numerous waterfalls, streams, and pools.

The zonal vegetation of Wugong Mountain belongs to the typical subtropical evergreen broad-leaved forest. The vegetation zone on the southern slope of the mountain range has strong tropical characteristics, while the northern slope has strong temperate characteristics. Vertical zonal vegetation types include mixed evergreen and deciduous broad-leaved forests, evergreen broad-leaved forests, coniferous forests, mountain dwarf forests, bamboo forests, and meadows. The forest coverage rate of Wugong Mountain reaches 88.1%, which is higher than the average level in China and even the world. There are 37 ancient trees and over 150 precious, rare, and endangered tree species preserved in Wugong Mountain [24]. In the current vegetation, there is a very rich variety of plant species. According to vascular plant statistics, there are 49 families, 114 genera, 403 species, 28 varieties, and 4 forms of ferns; 8 families, 22 genera, 29 species, and 2 varieties of gymnosperms (**Table 1**). There are more than 210 families and 1340 genera of angiosperms, accounting for about 60% of the total families and 38% of the total genera of angiosperms in China. *Ginkgo biloba* L is a relic plant from the Jurassic period of the Mesozoic era; *Cryptomeria fortunei* is a coniferous tree that continued from the Jurassic period to the present day. There is currently a species called *Cryptomeria fortunei*, which is distributed in primitive forests on Wugong Mountain. *Glyphostrobus penilis*, *Cephalotaxaceae*, and *Taxaceae*, which appeared as early as the Cretaceous 100 million years ago, have continued to this day and become a component of mixed coniferous and broad-leaved forests, or companion species in evergreen broad-leaved forests. *Taxus chinensis*, a shallow-rooted plant with an inconspicuous main root and well-developed lateral roots, is recognized as an endangered natural rare anti-cancer plant worldwide. It is an ancient tree species left over from the Quaternary glaciers and has a history of 2.5 million years on the Earth.

2.3 Social humanities

Wugong Mountain has a long history and a rich cultural heritage. Since the Han (202–220 BC) and Jin (266–420 AD) dynasties, it has been chosen by both Taoism and Buddhism as a blessed place for self-cultivation. During the Song and Ming dynasties, when incense was at its peak, there were more than 30 nunneries of arhat halls, buddhist temples, and taoist temples built in the south and north of the mountain.

Type	Family	Genus	Species	Variety	Form
Ferns	49	114	403	28	4
Gymnosperms	8	22	29	2	—

Table 1.
Statistics on the number of vascular plants in Wugong Mountain.

To this day, there is a continuous stream of worshippers of good men and women, as well as famous scholars who appreciate mountaineering, recite poetry, and write poems, leaving countless precious ink marks. After Xu Xiake, a great Ming Dynasty traveler and geographer, visited Wugong Mountain, leaving behind the timeless quatrain of "Thousand peaks and towering jade hairpins, the five ridges are comparable to Wugong Mountain" [25]. According to research, more than a hundred ancient poems, plaques, and articles praising Wugong Mountain have been left behind. The altar community of southern China at the Jinding of Wugong Mountain, with a history of over 1700 years, is known as a unique feature of China; there are currently ten ancient pagodas in Jiulong Mountain, all of which were built in the Ming Dynasty and are considered the only group of ancient pagodas in the province. After restoration, they can recreate the scenery of the past.

3. Problems with mountain meadow ecosystems

Mountain meadows are a rare landscape type in the composition of mountainous tourist attractions. For a long time, Pingxiang City, Anfu County, and Yichun City around Wugong Mountain have all focused their attention on the unique resources of Wugong Mountain in the process of resource development and promoting rapid local socio-economic development [26]. In 1985, Wugong Mountain was approved by the People's Government of Jiangxi Province as the first batch of provincial-level key scenic spots. In 2005, it was successfully declared as a national key scenic spot. With the completion of the high-speed railway from Shanghai to Ruili, the railway from Zhejiang to Jiangxi has increased in speed, and the expansion of the tourism road from Luxi to Wugong Mountain has been carried out, especially the construction of the road network in the scenic area and the opening of the two-level mountain climbing cableway, the range of tourist sources that come with admiration has significantly expanded. At present, the customer source market has expanded to the Nanchang, Changsha, and the border areas between Jiangxi and Hunan provinces. In the medium and long term, a tourism radiation area can be formed in the East China region centered around Shanghai, the Central China region centered around Wuhan, and the South China region centered around Guangzhou. Since 2004, the Wugong Mountain has received 450,000 visitors. Since 2008, the annual Wugong Mountain International Tent Festival has also attracted more and more tourists from all over the country to climb, explore, and camp [27]. It is worth noting that with the expansion of tourism scale and the extension of tourist travel range, the negative impact on various vegetation and special landscapes, including mountain meadows, is also increasing day by day. The arbitrary trampling by tourists and the discharge of waste have largely led to the degradation of meadow community composition and productivity, and the gradual fragmentation of about 4000 hm^2 of mountain meadows is becoming increasingly apparent [28].

Overcultivation and overgrazing are some of the main causes of ecological and environmental problems in mountain meadows. Excessive cultivation can lead to intensified soil erosion, which in turn affects the safety of water sources. Overgrazing can lead to meadow vegetation degradation, and reduced biodiversity, and thus affect the stability of ecosystems. Therefore, limiting excessive cultivation and grazing is one of the important measures to protect the ecological environment of mountain meadows. Soil erosion is another important aspect of the ecological environmental problem in mountainous meadows. In addition, factors such as climate change may also lead to intensified soil erosion [29]. Therefore, taking measures to prevent soil erosion is one of the important measures to protect the ecological environment of mountain meadows. The reduction of biodiversity is another important aspect of the ecological environment of mountain meadows [30]. Besides to the above factors, other human disturbances also may lead to a gradual decrease in biodiversity, and some species may be endangered. Therefore, taking measures to protect biodiversity is one of the important measures to protect the ecological environment of mountain meadows.

4. Research on mountain meadow ecosystem

4.1 Research on plant biodiversity

Originating from the 1940s, biodiversity has become a crucial research domain within ecology and global change, presently commanding international ecological attention. Based on the hierarchy of life systems, different scholars divided biodiversity into four levels: genetic diversity, species diversity, ecosystem diversity, and landscape diversity [31].

The variation of biodiversity along environmental gradients is an important research topic in this field. Wang et al. [32] research shows that species diversity and evenness are highest in mid-elevation areas, and species evenness and diversity first increase and then decrease with increasing elevation. Among the currently rare diverse personality traits are: α diversity index increases with elevation, and there is no clear pattern in the species richness index of coastal gradient changes. Therefore, it is necessary to study the biodiversity of various important mountains in nature to accumulate a foundation for the pattern of mountain diversity.

Studying population spatial patterns has always been the most important research field in community ecology. Community pattern analysis is the process of studying the mosaic structure between small communities and small community gaps, or between community patches and community patch gaps. It is an important means of studying plant community structure. At present, common spatial distribution patterns include three types: random distribution, clustered distribution, and uniform distribution. Some scholars believe that spatial distribution patterns should be divided into four types: random distribution, clustered distribution, uniform distribution, and mosaic distribution [33]. The community environment is composed of biotic and abiotic factors. The spatial pattern often reflects the relationship between its dynamic environment and individual survival and growth, as well as indicates the ecological adaptation strategies of plants.

Based on the survey data of 115 sample plots in the distribution area of Wugong Mountain meadow, in previous studies, we analyzed the characteristics and distribution types of vascular plant flora, plant community diversity, spatial distribution patterns of small communities, and human interference effects in the meadow [34]. Through field investigations and specimen identification in the research area over the past 2 years, it was found that within the survey elevation range of 1500 to 1900 m, the meadow plant resources in Wugong Mountain are relatively abundant, including 44 families, 90 genera, and 108 species of ferns, gymnosperms, and angiosperms. Among them, there are 6 families, 6 genera, and 6 species of ferns; 1 family, 2 genera, and 2 species of gymnosperms; 37 families, 82 genera, and 100 species of angiosperms (31 families, 63 genera, and 81 species of dicotyledonous plants; 6 families, 19 genera, and 19 species of monocotyledonous plants). Through investigation and analysis of plant community species at different elevations and slopes in the Wugong Mountain meadow, it was found that *Miscanthus* occupies a dominant position in the entire plant community, followed by the wild ancient grass as a subdominant species. In addition, there are some accompanying species such as *Hypericum ascyron* and *Selaginella*.

We analyzed the spatial distribution pattern of the surveyed plant communities and found that *Miscanthus*, as the main community, has a relatively large distribution area, with 23 types of small communities. The dominant and constructive species in each small community are different, and they all exist as separate small areas in the surrounding large communities. These small communities are mostly distributed on both sides or near the main walkway, with small areas, and some are sporadically embedded in the large communities, a part of them is distributed in an island-like pattern on a mountain slope and is not universal. They are all unique to large communities. Similarly, in the vertical distribution pattern, it can be observed that there are more small communities distributed at elevations of 1650 and 1700 m compared to other elevations, with only one community found at an elevation of 1900 m. As the elevation gradient increases, the α diversity index changes in a wavy pattern (**Table 2**). Analyzing the changing trends of the five indices of the β diversity index shows that as elevation increases, all indices exhibit irregular jumping changes, which may be related to changes in factors such as temperature, water, soil, and light at elevation and changes in the degree of human interference.

The main tourist trail and meadows on both sides of Wugong Mountain's main peak (Jinding) have been deeply affected by tourism activities, especially the main peak, which has been seriously disturbed by human activities, and even several exposed plots have appeared. In response to these phenomena, a study was carried out on the effects of human interference on the meadow of Wugong Mountain. From the main peak of Wugong Mountain at 1900 to 1500 m (the lower limit of the meadow), sample plots were selected from the edge on the same side of the main trail that was most affected by tourism interference. A small sample plot of 1 m x 1 m was set up every 50 m above sea level, with each sample plot being 0 m, 4 m, 8 m, and 12 m away from the main track. Thus, four parallel transects were formed along the main track, totaling 36 small plots. During the investigation, record the plant name, abundance,

Elevation(m)	Species number	Shannon-wiener index	Simpson index	Pielou index
1500	45	3.455	0.9577	0.8878
1550	34	3.085	0.9425	0.9162
1600	50	3.6028	0.962	0.9032
1650	46	3.8392	0.9714	0.9972
1700	40	3.6424	0.9674	0.9745
1750	45	3.6394	0.967	0.9934
1800	31	2.7586	0.9176	0.868
1850	25	2.5256	0.8911	0.8577
1900	26	3.2981	0.9527	0.8941

Table 2.
The changes of plant community diversity in different elevation gradient.

coverage, frequency, and height of each small plot, as well as environmental factors such as longitude, latitude, and slope.

Tourism activities have had an impact on the community structure and species composition of the Wugong Mountain meadow to a certain extent. The farther away from the main walkway, the greater the total coverage of the transect, and the smaller the degree of tourism interference. In addition, the composition and structure of plant species in the meadow have also changed. Along the main trail, plants that were originally dominant in the meadow, such as *Miscanthus* and *Arundinella hirta*, have become companion species due to poor anti-interference ability. However, plants that were originally companion species, such as *Gerbera delavayi* Franch, *Elsholtzia stachyodes* (Link) C. Y. Wu, and *Senecio latouchei* J. F. Jeffrey, have become dominant species. These plants have strong competitiveness, wide adaptability, and certain anti-interference abilities under the interference of human tourism activities. This may be due to interference that can regulate the ratio of plant species and plant numbers between local and non-local species, thereby affecting the species composition of plant communities. With the vigorous development of the tourism industry, frequent human activities have had a certain impact on the vegetation growth and plant community structure of the Wugong Mountain meadow.

The species richness index of Wugong Mountain meadow shows a decreasing trend with the decrease of tourism disturbance. Research has shown that a significant decrease in species richness index can be observed from the first parallel transect to the second, indicating that tourism interference causes changes in the internal composition structure of plant communities. However, there is no significant change in the richness indexes in the second parallel transect to the fourth, and the intensity of tourism interference does not have a substantial impact on the structure of plant communities. The first parallel is located at the edge of the main trail, which is most disturbed by tourism activities. Human activities have led to the invasion of some companion plants, damaging the original plant community and increasing the species richness index at the edge of the main trail.

The higher the elevation areas, the lower the coverage and the greater the degree of grazing disturbance. The species richness index shows an increasing trend with the degree of interference, and both the diversity index and evenness index show a trend of "first increasing, then decreasing, and then increasing". The evenness index is highest for mild interference and lowest for moderate interference; the diversity index of undisturbed plants is the lowest, while the diversity index of heavily disturbed plants is the highest, indicating that the strength of grazing disturbance has a certain degree of impact on plant community diversity.

4.2 Research on soil characteristics of mountain meadows

Soil is an important component of the Earth's biosphere and the material foundation on which humans and other animals and plants rely for survival. Soil permeability is one of the important factors affecting soil erosion and an important indicator reflecting the function of soil water conservation and regulation. As a sensitive indicator for stabilizing ecosystems and monitoring changes in soil quality, the diversity of soil microorganisms has played a huge role in evaluating ecosystems and maintaining ecological balance. Therefore, more and more scholars are turning their attention to the research and protection of soil microbial diversity. Soil microorganisms are not only an important component of soil, but also the main drivers of soil nutrient cycling. Changes in soil microbial communities can to some extent reflect the trend of soil quality changes [35]. Among soil microorganisms, bacteria and fungi have the most types and quantities. Bacteria and fungi play an extremely important role in soil nutrient cycling, organic matter formation and decomposition, soil structural composition formation, ecological environment improvement, plant growth and development, and crop pest control [36]. Therefore, research on the characteristics of microorganisms dominated by bacteria and fungi has received widespread attention. Soil nutrients are essential nutrients for plant growth provided by the soil. Soil available nitrogen (AN), available phosphorus (AP), and available potassium (AK) nutrients are the direct sources of the three essential basic elements for crop growth and development [37–39]. Elevation is an important ecological factor that affects crop layout and growth and development, and also has a significant impact on the content of various nutrients in the soil [40, 41]. Natural meadow vegetation is the largest green vegetation layer on the Earth's land surface and an important component of terrestrial ecosystems [42, 43].

The Wugong Mountain meadow in Jiangxi province is a typical representative of subtropical mountain meadows, which have typicality and specificity in the vertical vegetation spectrum of east China due to its wide area and low distribution benchmark elevation. Due to the sharp increase in the number of tourists, the meadow ecosystem has undergone varying degrees of degradation due to human trampling, plant populations have evolved, age structures have changed, biodiversity has decreased, and productivity has declined. Meadow plants exhibit characteristics such as dwarfism, inferiority, and sparsity, some of which even become exposed to the surface, leading to a decrease in fragile ecosystem functions and a weakened ability to recover. Li et al. [44] studied the hyperspectral characteristics of several vegetation in the region and found that the spectral reflectance of the vegetation was as follows: *Carex chinensis* > Arundinella anomala>*Miscanthus sinensis* > *Sinarinaria nitida* > *Fimbristylis wukunnshanensis*; Zhao et al. [21] research showed that the total amount of inorganic phosphorus in meadow soil significantly increased with increasing elevation; Deng et al. [45] believe that as the elevation increases, there is no significant change in the content of available Fe, Cu, Zn, B, and other elements in the meadow soil. However, as the degree of degradation increases, the content of available Cu shows an enrichment phenomenon; Yuan et al. [46] found that the alkaline nitrogen content in the 0–20 cm soil layer of the meadow in the region was higher than that in the 20–40 cm soil layer.

We also found that there is variation in the soil water conservation function of meadows at different elevations in Wugong Mountain, but the variation characteristics are not completely consistent with the changes in elevation [47]. The best water conservation capacity is observed within a range of 1600 m, followed by 1800 and 1700 m, while the water conservation capacity is weak within a range of 1900 m. The soil permeability ranges from 1600 to 1800 m and gradually decreases with increasing elevation, while it increases again in the high-elevation range of 1900 m. As meadow degradation intensifies, soil permeability gradually decreases. The universal empirical model can better fit the soil infiltration process in this study area. The variation characteristics of soil pH show an increase with elevation, an increase in soil activity with the degree of meadow degradation, and a decrease in pH value; soil organic matter remains consistent across varying elevations and degraded meadows; soil potassium concentration rises with elevation, whereas both total phosphate and available phosphate quantities augment with meadow degradation; available sulfur diminishes with elevation ascent and goes up with meadow degeneration; available manganese quantity exhibits a drop with elevation and degradation increment.

At different elevation ranges, the variation characteristics of soil microbial distribution mainly come from the 1900 or 1800 m range of high-elevation areas, while the variation characteristics of soil microbial distribution in mountain meadows with different degrees of degradation mainly come from areas with severe meadow degradation. As the elevation increases and the degree of meadow degradation intensifies, the biomass per unit area of meadows generally shows a decreasing trend. There is a significant positive correlation between bacterial diversity index and leaf biomass, while the number of bacterial OUTs (Operational Taxonomic Units) is positively and negatively correlated with leaf biomass ratio and root biomass ratio, respectively. There is a certain negative correlation between *Cyanobacteria, Archaea*, Nitrifying spirochetes, and biomass. The *Saccharomycota* is positively correlated with total vegetation biomass, while the *Chlamydomonas* is negatively correlated with aboveground vegetation biomass. Soil catalase is positively correlated with both underground and total biomass of vegetation, while sucrase is negatively correlated with aboveground biomass. There is a negative correlation between total potassium and available phosphorus in soil and aboveground biomass of vegetation [18].

4.3 Response of mountain meadow ecosystem to climate change

Climate change is one of the important issues facing the world today, among which rising atmospheric CO_2 concentration and temperature are the main characteristics of future climate change. These changes have a significant impact on the soil nitrogen mineralization process, which in turn affects crop growth and soil quality. After crop residues enter the soil, the content of available nitrogen in the soil is influenced by the dynamic changes of microbial fixation and organic nitrogen mineralization. On the one hand, under high atmospheric CO_2 concentration, the C/N content of crop residues is higher. Soil microorganisms not only obtain rich carbon sources but also increase the fixation of inorganic nitrogen in the soil. Although microorganisms may promote the mineralization of soil organic nitrogen to utilize more inorganic nitrogen, it is manifested overall as a net biological fixation of mineral nitrogen, resulting in a decrease in effective nitrogen content in the soil. On the other hand, C/N in soil decreases with organic decomposition and CO_2 release, and the rate of nitrogen fixation by microorganisms gradually decreases compared to the mineralization rate of organic nitrogen, resulting in net mineralization of organic nitrogen. Therefore, under the condition of elevated CO_2, although crop residues strengthen microbial nitrogen fixation in the short term, they may ultimately promote soil organic nitrogen mineralization, increase the available nitrogen content in the soil, and facilitate crop absorption and utilization. Climate change has a significant impact on soil nitrogen mineralization, mainly manifested in the increase of atmospheric CO_2 concentration and temperature, as well as the impact of crop residues on soil nitrogen mineralization [48].

Nitrogen is an important nutrient element in living organisms, one of the main components of proteins, and a key factor in regulating the production, structure, and function of terrestrial ecosystems. It can limit the primary and secondary productivity of communities. The disruption of the balance of nitrogen cycling will inevitably affect the ecological balance of other important physical, chemical, and biological properties, leading to changes in the global environment. The core problem of meadow degradation is soil degradation, which has a certain buffering and complexity. Nitrogen is the most abundant element in the atmosphere and an essential element for plant growth and development. It is crucial in the global carbon and nitrogen cycle. The nitrogen absorbed by plant growth and development mainly comes from the soil; therefore, the chemical form and transformation of soil nitrogen restrict the absorption and utilization of nitrogen by plants. The detection of soil nitrogen usually selects total nitrogen (TN), alkaline nitrogen (AN), nitrate nitrogen (NO3−N), and ammonium nitrogen (NH4 + −N).

Our research starts with the response of the soil nitrogen pool in the Wugong Mountain meadow to ecosystem degradation and simulated warming, exploring the changes in soil physicochemical properties with meadow degradation, as well as the impact of climate change on meadows. With nitrogen as the breakthrough point, we accurately grasp the impact of meadow degradation on soil nutrients. There are many reasons that can cause meadow degradation, such as overgrazing, climate change, outbreaks of herbivorous small mammal populations, and interference from tourism behavior. The degradation of meadows leads to soil degradation, and the two are mutually causal and cyclical. The mountainous meadow ecosystem has important ecological functions such as improving soil, regulating climate, purifying air, and conserving water sources and its nitrogen storage is enormous. However, due to excessive tourism development, the degradation of Wugong Mountain meadow is becoming increasingly serious. This not only leads to the destruction of soil structure, soil fertility, and vegetation diversity in the meadow ecosystem but also causes significant changes in the composition and structure of its meadow vegetation; the regional ecological environment is facing severe challenges. Therefore, maintaining the stability of meadow ecosystems is of great significance for the development of the regional ecological environment and nitrogen balance. A study on nitrogen mineralization in the meadow soil of Wugong Mountain found that under the same temperature cultivation, as degradation intensifies, the net nitrogen mineralization rate will gradually increase. Under the same elevation and degree of degradation, the net nitrogen mineralization rate of soil will significantly increase from 15 to 25°C ($P < 0.05$), and slightly decrease from 25 to 35°C, with an overall performance of 25°C > 35°C > 15°C. Therefore, it is predicted that the peak of soil net mineralization in Wugong Mountain meadow may occur between 25 and 30°C [22, 48].

5. Research on the protection and restoration of mountain ecosystems

The protection of mountain ecosystems mainly relies on legal, policy, scientific research, and educational means to reduce the damage of human activities to the ecosystem, promote the restoration of the ecosystem, and protect biodiversity. Governments of various countries protect mountain ecosystems by formulating relevant laws and regulations. For example, regulations such as the Regulations on Nature Reserves of the People's Republic of China and the Wildlife Protection Law of the People's Republic of China have played an important role in protecting mountain ecosystems and wildlife. The government protects mountain ecosystems by formulating various policies and plans. The Chinese Government has implemented policies such as natural forest protection projects and land conversion projects to protect mountain forest ecosystems. Chinese scientists want to understand the ecological processes and biodiversity of mountain ecosystems through scientific research, they have discovered many unique plant and animal species on the Qinghai Tibet Plateau, which provided a scientific basis for protecting these species. The Chinese government want to enhance public awareness and awareness of mountain ecosystem protection through educational and promotional activities, some education departments has added content related to nature conservation in primary and secondary school textbooks to enhance young people's awareness of nature conservation.

The restoration of mountain ecosystems is mainly promoted through vegetation restoration, soil restoration, and water resource management to promote the restoration of ecosystems and improve the ecological environment. Through measures such as aforestation and returning farmland to forests, the vegetation of mountain ecosystems can be restored, and the biodiversity and stability of the ecosystem can be improved. China's Grain for Green project improves the stability and disaster resistance of mountain ecosystems by gradually restoring steep slope farmland to forest vegetation. By taking physical, chemical, and biological measures to repair damaged soil, the quality and productivity of the soil can be improved. Some regions in China have adopted measures such as returning straw to the field and applying organic fertilizers to improve soil structure and enhance soil fertility. By rational allocation and effective management of water resources, we can ensure the normal operation of mountain ecosystems and the full play of ecological service functions. Other regions in China have adopted measures such as rainwater collection and water-saving irrigation to improve the efficiency and management level of water resources.

The Wugong Mountain Meadow in Jiangxi Province is a typical representative of subtropical mountain meadows. Currently, human interference and excessive tourism development have led to serious degradation and fragmentation of the fragile meadow ecosystem in the region. It is urgent to carry out reasonable vegetation restoration to prevent further deterioration of the ecological environment [49]. Bai et al. [50] summarized the theoretical and practical work of early degraded ecosystem restoration and reconstruction, and believed that plant measures are the primary step and preferred solution in restoration measures. Yang et al. [51] used ryegrass (*Lolium perenne* L.), *Poa annua* L., and creeping purple fescue (*Festuca rubra* L.) as sowing grass seeds, plant paper, and non-woven fabric as auxiliary materials, and conducted fertilization treatment experiments. It was found that non-woven fabric and plant paper have similar auxiliary effects on vegetation growth, and an increase in fertilization will actually reduce the emergence rate. Ge et al. [52] restored the degraded areas of alpine meadows in Maqu County, Gansu Province through measures such as fertilization, supplementary sowing, and soil plowing, and believed that several measures were effective. Yang et al. [53] conducted a study on the vegetation restoration of the slopes of the Yuexi to Wuhan expressway in the western Anhui region and found that using plant fiber blankets as auxiliary materials has better effects.

The research team led by the author has currently conducted research on soil nutrient distribution [17, 20], vegetation hyperspectral characteristics [44], and mountain meadow tourism marketing [54] in the Wugong Mountain meadow, enriching the scientific research achievements of the vegetation ecosystem in the region. The degradation of meadows can generally be attributed to two major factors: natural and human factors [43]. The vegetation degradation of Wugong Mountain meadows is mainly caused by tourism development and human trampling. How to adapt to the environmental requirements of the region and have good tolerance to the trampling behavior of tourists, while quickly repairing the current ecological degradation situation, is an important issue faced by scientific research and business planning personnel. We take the severely degraded area of Wugong Mountain meadow as the research object and analyze the vegetation growth effect and adaptability characteristics under different treatment methods by sowing different grass species, covering different auxiliary materials, and conducting turf transplantation and enclosure. We explore excellent measures suitable for vegetation restoration in the area, providing theoretical guidance for the ecological restoration of degraded meadows, providing a reference basis for the management planning and sustainable development of subtropical mountain meadows. Among several vegetation restoration measures, direct sowing of awns has the best growth effect, followed by dogtooth root, tall fescue, ryegrass, thrush grass, and enclosed vegetation have poor growth effects. Covering non-woven fabric as an auxiliary measure has a better vegetation growth effect than covering grass curtains and non-covering measures. Grass transplantation has the best growth effect among all vegetation restoration measures, and the grass transplantation effect of local grass species such as *Miscanthus*, *Carex chinensis*, and *Fimbristylis wukungshanensis* is better than that of imported grass species such as *Cynodon dactylon*. In the process of vegetation restoration, it is recommended to use measures such as grass transplantation based on the premise that it does not cause significant damage to the original vegetation environment; in addition, it is recommended to use *Miscanthus* as the main grass species for restoration, *Cynodon dactylon* as an auxiliary grass species, and cover it with non-woven fabric for vegetation restoration [55].

6. Development and utilization of sustainable mountain ecosystems

The development of mountain ecosystems mainly utilizes resources through sustainable ecotourism, agriculture, and forestry to promote economic development and reduce poverty [56]. Some local governments and tourism companies utilize the natural landscapes and cultural heritage resources of mountainous ecosystems to develop the ecotourism industry and provide support for local economic development. China's Jiuzhaigou Valley Scenic and Historic Interest Area, Zhangjiajie, and other tourist attractions have attracted a large number of tourists to come for sightseeing and vacation by using their unique natural landscape and cultural heritage resources. There are also some local governments and business operators utilize the agricultural resources of mountainous ecosystems to develop characteristic agriculture and improve the income and living standards of farmers. The province of Guizhou and Yunnan in China have utilized their abundant agricultural resources to develop characteristic agricultural industries such as tea and flowers. If there are mountains within the administrative area, government departments and tourism enterprises Utilize the forestry resources of mountainous ecosystems to develop the forestry industry and provide support for local economic development. China's Changbai Mountain and Huang Mountain have used their rich forestry resources to develop forestry industries such as wood processing and undergrowth economy [57].

The tourism industry is affected by the environment. In a certain sense, it can be said that tourism is developed based on the environment. Only a beautiful environment can make people feel relaxed and happy, leaving them lingering and immersed in the beautiful atmosphere. At the same time, it can cultivate sentiments, stimulate enthusiasm, achieve spiritual satisfaction, and discover, enjoy, and spread beauty in the beautiful scenery. On the contrary, if there is no high-quality environment or the environment is damaged, it cannot attract tourists to travel. Protecting the environment is the fundamental guarantee for the sustainable development of tourism, and it is the most basic condition for the sustainable development of a country or region's tourism industry. At the same time, good tourism development can also promote the protection of the tourism environment and provide local environmental investment, which complement each other [54].

However, in recent years, with the rapid development of the tourism industry and the arrival of the era of mass tourism, a large number of tourists have flooded into tourist destinations, putting enormous pressure on the sustainable development of tourist destinations. Some problems such as resource destruction, declining landscape quality, and environmental pollution are becoming increasingly serious, mainly manifested in (1) endogenous destruction, which is the damage caused by tourism activities themselves, there are mainly constructive damages to tourist attractions and pollution of the tourism environment caused by household waste. (2) Exogenous damage, namely the damage caused by economic activities in the surrounding areas, mainly includes pollution of the tourism environment caused by waste gases and water, and unreasonable layout of infrastructure, which damages the tourism environment. However, the traditional tourism environment management in our country still remains at a low level due to the emergence of diverse policies and chaotic management. (3) Without new breakthroughs, the sustainable development of scenic spots is greatly threatened. Based on this, tourist attractions must seek more suitable sustainable development strategies and scenic area tourism environment management methods. Experts in the tourism and natural ecology fields are also actively considering how to protect tourism resources, prevent environmental pollution and damage, and achieve an organic unity of economic, social, and environmental benefits [58].

Wugong Mountain is a mountainous scenic area with natural scenery as its main feature. The mountainous meadows are a rare landscape type in the composition of mountainous tourist attractions. By comprehensive and systematic investigation, analysis, and evaluation of the tourism resources, a survey and analysis were conducted on the recreational behavior of tourists; previous researchers conducted an analysis and evaluation of various recreational activities' impact on the tourism environment factors of Wugong Mountain's scenic area. We considered tourist experience and resource conservation aspects to construct a range of mountain grassland recreation opportunities, calculate the tourism environmental carrying capacity, and offer pertinent suggestions for environment management, including national park establishment, waste management, greenway development, and ancient tree cultural restoration [57, 59–61].

6.1 Analysis and evaluation of tourism resources

The Wugong Mountain Scenic Area has a wide variety of tourism resources with complex structures and abundant individual resources. The overall evaluation score for tourism resources is 80 points, including 9 fifth-level tourism resources. The score for typical landscapes (mountain meadow landscapes, celestial wonders, and ancient tree landscapes) is 85 points, indicating a high score and high overall development value.

6.2 Analysis of recreational behavior characteristics of tourists

Based on first-hand data obtained from questionnaire surveys and in-depth interviews, this study focuses on mountain tourists and analyzes the general characteristics of the Wugong Mountain recreational market and tourist recreational behavior. The results indicate that the recreational behavior of tourists in Wugong Mountain has the following obvious characteristics:

1. In terms of demographic characteristics, the gender structure is dominated by male tourists, with middle-aged and young tourists as the main group. Among

them, the proportion of male and female tourists is 60.7 and 39.3%, and tourists between the ages of 18 and 40 account for 89.3%. In terms of occupation and education, student tourists account for 38.2%. In addition, company employees and enterprises and institutions account for 17.1 and 13.2%, respectively, with a higher proportion. From the perspective of educational background, tourists have the characteristic of high education, with 70.4% having a bachelor's degree and 5.3% having a graduate degree. Therefore, young students also as the main group, with high educational qualifications. However, the proportion of high-end outdoor recreational tourists is relatively small.

2. In terms of the spatial structure of the customer source market, the market structure is mainly characterized by short-range customers, supplemented by remote customers, and a combination of short-range and remote customers.

3. In terms of general recreational behavior characteristics, there is an inherent consistency between tourist motivation and activity preferences, and they exhibit a preference for recreational activities in mountain meadows and backpacking characteristics.

4. In terms of recreational experience evaluation, tourists have higher satisfaction and willingness to revisit, and there is also an inherent consistency between the two, but their satisfaction with the meadow tourism environment is relatively low. Tourist satisfaction is an important indicator of experience evaluation, which can reflect their perceptions and evaluation opinions on tourism products and services. The Likert scale was used to analyze visitor satisfaction in Wugong Mountain's recreational environment. Five attitudes, namely very dissatisfied, dissatisfied, basically satisfied, satisfied, and very satisfied, are expressed in the form of 1–5 points. After statistical analysis, the overall satisfaction of tourists with the recreational environment of Wugong Mountain is 3.45, indicating that tourists are generally satisfied with the recreational environment of Wugong Mountain at present. This is basically in line with the recreational expectations of tourists. 38.6% of tourists believe that the recreational experience of Wugong Mountain is "mostly what I expected" and 37.3% of tourists feel that the recreational experience is "good, but I hope there are still some aspects that can be done better". Generally speaking, tourists' satisfaction directly affects their rate of revisiting and their willingness to revisit. 80.5% of first-time tourists come to Wugong Mountain, of which 62.3 and 19.7% are first-time tourists who make "possible" and "certain" choices to participate in outdoor recreation in Wugong Mountain again. The inherent consistency between tourist satisfaction and willingness to revisit indicates that the recreational environment and projects provided by Wugong Mountain Scenic Area are generally recognized by tourists, but 18% of tourists choose "unlikely" to revisit.

6.3 Analysis of the impact of the recreational behavior of tourists in Wugong Mountain on tourism environment

The impact of recreational behavior of tourists in Wugong Mountain on the meadow environment was analyzed from the aspects of trampling behavior, camping activities, tourism waste, and hiking, mainly using on-site experiments and investigations. The results indicate that:

1. In the experimental state, the height, coverage, number of plant clusters, and dominant species of the meadow near the Wugong Mountain plot all increased with the increase of distance from the main road, while the number of accompanying species did not change significantly. As the intensity of trampling interference increases, the average height and coverage of the meadow plant community decrease, and the horizontal structure of the community undergoes significant changes. However, with the passage of time after trampling, both the community coverage and height show a trend of recovery, almost approaching the pre-trampling standard (except for 700 trampling intensities). Under the interference of trampling experiments, as the intensity of trampling interference increases, the number of dominant and associated species also increases. Under high-intensity trampling, there were significant changes in the combination of dominant and associated species compared to the control group, but the combination of dominant and associated species tended to stabilize; the impact of over 200 trampling intensities on meadow vegetation diversity still exists 2 months after the end. Moreover, the meadow plant community exhibits response characteristics (resistance and recovery) under trampling interference. When the trampling intensity reaches 1325 times, the relative coverage decreases by nearly 50%. The required trampling intensity for every 50% increase in relative coverage should continue to increase, and there is a non-linear relationship between trampling intensity and flat coverage. Under trampling interference, there are two response thresholds for vegetation coverage (89.7 ± 1.2% ~ 78.3 ± 0.7%, 78.3 ± 0.7% ~ 62.1 ~ 0.9%), corresponding to trampling intensity of 313 and 683 times, respectively. When the trampling intensity is below 313 times, the decrease in vegetation coverage is relatively slow; when the trampling intensity is between 313 and 683 times, the vegetation coverage decreases at a relatively stable rate; when it exceeds 683 times, the vegetation coverage decreases at a faster rate. Based on the decrease in meadow community coverage after 1 week of trampling, the meadow community coverage recovered by 29.7, 58.96, and 78.5% after 2 weeks, 1 month, and 2 months of trampling, respectively.

2. In natural conditions, tourist camping activities have a significant impact on the average height and number of plants in the meadow, both showing a decreasing trend with activity intensity. The height increased from 31 to 5 cm, and the number of plants increased from 252 to 94. The numerical variation of the impact on soil bulk density is more significant, and the soil is more compact, which affects the extension and growth of meadow roots and the permeability of water. However, there is not much change in pH value after the holiday, and camping has no significant impact on soil acidity and alkalinity. The numerical changes in the impact on soil bulk density are also more significant, and the soil is more compact, affecting the extension and growth of meadow roots and water permeability.

3. In the natural state, in the survey of tourists' preferences for existing road choices, 67.77% of the upward trend and 46.28% of the downward trend choose bare land. The maximum and minimum values of road width among the 5 abnormal path widths have increased, indicating that the abnormal paths have been widened due to the trampling of tourists. The meadow vegetation and soil near the road are significantly affected by the trampling of tourists. The height

of the meadow vegetation on both sides of the road increases gradually from 1 to 5 m, and the types and associated species of meadow vegetation at 1 m are more affected by human activities than at 5 m.

4. The tourism waste in Wugong Mountain Scenic Area presents a large amount, wide coverage, and concentrated types, while the existing disposal methods are still simple and crude, which has a significant negative impact on the environment, image, and tourist satisfaction of the scenic area.

6.4 Research on the carrying capacity of tourism environment

Twenty evaluation indicators were selected from three aspects: elasticity, support, and resistance, and a bearing capacity evaluation index system for the recreational environment of Wugong Mountain meadow was constructed. The threshold range of each indicator and bearing capacity index was determined, and a corresponding bearing capacity index model was constructed.

The comprehensive index (LO) of the carrying capacity of the recreational environment in the Wugong mountain meadow during peak season is 0.4263, and it presents a carrying pattern of "high elasticity-low support-low resistance". The elasticity index (LA), support index (LB), and resistance index (LC) are 0.0644, 0.2066, and 0.1553, respectively. In the off-season, its comprehensive bearing capacity index (LO) is 0.5661, showing a bearing pattern of "high elasticity full support high resistance". The elasticity index (LA), support index (LB), and resistance index (LC) are 0.0656, 0.2513, and 0.2492, respectively.

From a single indicator perspective, the recreational environment of the Wugong Mountain meadow is significantly overloaded during peak season, with 65% of indicators in an overloaded state, 25% in a fully loaded state, and only 10% in a low-loaded state. During the off-season, the indicator carrying capacity is diverse, with 35% in the "overloaded" and 35% in the "fully loaded" states, and 30% in the "low-loaded" state.

6.5 A tourism environment management strategy for Wugong Mountain based on tourist recreation behavior

Based on the laws and characteristics of tourist recreational behavior, this study attempts to strengthen and improve the management of the tourism environment in Wugong Mountain through strategies such as introducing national park management system pilot construction, constructing recreational opportunity spectrum, garbage management, greenway construction, and ancient tree culture production.

1. Pilot construction of national park management system.

Wugong Mountain stretches for 120 km with a total area of over 260 km^2. It is a complete ecosystem that integrates cultural and natural landscapes, fully in compliance with the national park standards recognized by the International Union for Conservation of Natural Resources (IUCN). Given the current situation of the management committees for one mountain and three scenic spots, and against the backdrop of the country actively promoting the pilot construction of national parks, it is necessary and appropriate for Wugong Mountain to promote the pilot construction of the national park management system.

2. Building a recreation opportunity spectrum for Wugong Mountain.

Based on the characteristics of the recreational environment of the Wugong Mountain meadow and the comprehensive consideration of the environmental recreational possibilities of the mountain meadow tourist area, and based on the questionnaire survey of the recreational activities and experiences expected by tourists in three different environmental types, and the analysis of the importance and preference of the comprehensive recreational environmental factors, a mountain meadow ecological recreational opportunity spectrum is determined based on three aspects: environmental type, recreational activities, and recreational experiences.

3. Wugong Mountain tourism garbage management model.

The reason for the problem of tourism waste is the result of the joint action of relevant stakeholders, and the existing governance methods mainly rely on technical and management level governance led by a single entity, which has not been effective. Stakeholder theory should be introduced to construct a tourism waste management model for mountainous scenic areas, which involves a collaborative approach of multiple stakeholders to implement full process management from source control before tourism waste generation to maintenance after cleaning, achieving effective management of tourism waste in mountainous scenic areas.

4. Wugong Mountain Greenway construction.

The Wugong Mountain Greenway, also known as the Wugong Mountain Ecological Trail or Scenic Trail, has important practical significance for achieving sustainable development of the Wugong Mountain meadow. It helps to improve the ecological system, increase cultural charm, improve reception facilities, and enhance safety management. The construction of greenways in the Wugong Mountain meadow should choose ecological, cultural, humane, and safe approaches.

5. The production mechanism of ancient tree culture in Wugong Mountain.

There are over 100 ancient trees and 37 varieties, which are relatively concentrated. Firstly, in the inheritance and protection of ancient tree culture, we should actively maintain the individuality of ancient trees and selectively inherit and utilize ancient tree culture; secondly, efforts should be made to shape the image of ancient trees and reconstruct their landscapes, striving to amplify the cultural value of the trees themselves and expand the cultural space for their survival; thirdly, cultural transplantation should be carried out on unnamed and storyless ancient trees, borrowing foreign cultural resources to make them important tourism resources.

7. Conclusion

The Wugong Mountain meadow ecosystem is a typical alpine meadow in southern China, with important ecological value and unique tourism resources. The core of protecting and restoring the meadow ecosystem of Wugong Mountain lies in establishing and implementing strict management systems. In the process of protecting and restoring the meadow ecosystem of Wugong Mountain, we need to pay attention to the balance between ecological protection, resource utilization, and economic development. Only through scientific management and protection measures can the sustainable utilization and development of mountain meadow ecosystems be achieved.

Protecting, developing, and restoring mountain ecosystems are a long-term and arduous task that requires the joint efforts of the government, enterprises, and the public. In recent years, researchers have achieved rich research results in this field. However, further research is still needed in this field, including improving protection measures, development models, restoration technologies, and policy regulations, in order to better achieve sustainable development of mountain ecosystems. In addition, it is necessary to strengthen scientific research and technical support to provide a theoretical basis for the protection, development, and restoration of mountain ecosystems.

Conflict of interest

The authors declare no conflict of interest.

Author details

Zhi Li[1,2,3], Yanmei Wang[1,2,3], Xiaodong Geng[1,2,3], Qifei Cai[1,2,3], Zhen Liu[1,2,3], Ling Zhang[4*] and Xiaomin Guo[4]

1 College of Forestry, Henan Agricultural University, Zhengzhou, China

2 Henan Province Engineering Technology Research Center for Idesia, Zhengzhou, China

3 National Forestry and Grassland Administration Key Laboratory for Central Plains Forest Resources Cultivation, Zhengzhou, China

4 College of Forestry, Jiangxi Agricultural University, Nanchang, China

*Address all correspondence to: lingzhang09@126.com

References

[1] Erich T et al. Shaping the European Alps: Trends in landscape patterns, biodiversity and ecosystem services. Catena. 2024;**235**:107607

[2] Han Q et al. Identifying the relationships between landscape pattern and ecosystem service value from a spatiotemporal variation perspective in a Mountain–Hill–plain region. Forests. 2023;**14**:2446

[3] Olga K et al. Distribution of Cronartium x flexili, an interspecific hybrid of two fungal tree rust pathogens, in subalpine forest ecosystems of western USA. Fungal Biology. 2024;**128**:1578-1589

[4] Gong C et al. Spatiotemporal change and drivers of ecosystem quality in the loess plateau based on RSEI: A case study of Shanxi, China. Ecological Indicators. 2023;**155**:111060

[5] Yu F et al. How do mountain ecosystem services respond to changes in vegetation and climate? An evidence from the Qinling Mountains, China. Ecological Indicators. 2023;**154**:110922

[6] Zhang Z-S et al. Which is more important in stabilizing soil organic carbon in mountain ecosystems: Molecular features, mineral protection, soil nutrients, or elevation? Catena. 2023;**232**:107395

[7] Jin T et al. Spatiotemporal evolution of ecosystem service value and topographic gradient effect in the Da-Xiao Liangshan Mountains in Sichuan Province, China. Journal of Mountain Science. 2023;**20**:2344-2357

[8] Li J-H et al. Ecosystems face the risk of ecological deficits in the southern foothills of the Himalayas. Ecological Indicators. 2024;**158**:111637

[9] Yang Y et al. Geographical, climatic, and soil factors control the altitudinal pattern of rhizosphere microbial diversity and its driving effect on root zone soil multifunctionality in mountain ecosystems. The Science of the Total Environment. 2023;**904**:166932

[10] Qu H-J et al. Exploring coordinated development and its driving factors between carbon emission and ecosystem health in the southern hilly and mountainous region of China. Frontiers in Environmental Science. 2023;**11**:1289531

[11] Zhang Z-C et al. Degradation leads to dramatic decrease in topsoil but not subsoil root biomass in an alpine meadow on the Tibetan plateau, China. Journal of Arid Land. 2020;**12**(5):806-818

[12] Rana S et al. Soil quality assessment in tourism-disturbed Subtropical Mountain meadow areas of Wugong Mountain, central Southeast China. Life. 2022;**12**:1136

[13] Rana S et al. Evaluation of soil and water conservation function in the Wugong mountain meadow based on the comprehensive index method. Heliyon. 2022;**8**:e11867

[14] Jiang L et al. Alpine meadow restorations by non-dominant species increased soil nitrogen transformation rates but decreased their sensitivity to warming. Journalof Soilsand Sediments. 2017;**17**:2329-2337

[15] Li Z et al. Response of soil sulfur availability to elevation and degradation in the Wugong Mountain meadow, China. Plant, Soil and Environment. 2017;**63**:250-256

[16] Li Z et al. Distribution characteristics of soil available nutrients in subtropical mountainous meadow: A case study of Wugong Mountain in Jiangxi Province. Southwest China Journal of Agricultural Sciences. 2017;**30**(10):2308-2314

[17] Zhao X-R et al. Vertical distribution character of soil inorganic phosphorus in mountain meadow system of WuGong Mountain. Pratacultural Science. 2014;**31**(9):1610-1617

[18] Li Z. Variation of Soil Microbial Characteristics and its Influencing Factors in Wugong Mountain Meadow. Nanchang: Jiangxi Agricultural University; 2017

[19] Hou X-J et al. Vertical distribution characteristics of Miscanthus sinensis roots and its relationship to soil nutrients in the Wugong Mountains. Pratacultural Science. 2017;**34**(12):2428-2436

[20] Yuan Z-Y et al. Canonical correlation analysis between soil active organic carbon and soil nutrient in different vegetation small communities of Wugong mountain alpine meadow. Journal of Central South University of Forestry & Technology. 2016;**36**(2):84-90

[21] Zhao X-R et al. Spatial distribution of soil P and its correlation with soil acidity in mountain meadow of Wugong Mountain. Acta Agriculturae Universitatis Jiangxiensis. 2013;**35**(6):1223-1228

[22] Zhao Z-W et al. Change of soil nitrogen and physical property in different degradation stages of meadow on Wugong Mountain. Guangdong Agricultural Science. 2015;**42**(13):71-75

[23] Liu J et al. Zircon LA-ICPMS U-Pb dating of Hukeng granite in Wugongshan area, Jiangxi Province and its geochemical characteristics. Acta Petrologica Sinica. 2008;**24**(8):1813-1822

[24] Liao Q-S et al. The diversity and protection of the rare and endangered endemic plant species in Wugong Mountain. Journal of Pingxiang College. 2008;**25**(3):79-83

[25] Zhao L-G. Research on the Construction of a Tourism Interpretation System for Wugong Mountain Meadows. Nanchang: Jiangxi Agricultural University; 2015

[26] Wang J-L. The characteristics of sports tourism resources in Wugong Mountain and its development. Journal of Pingxiang College. 2013;**30**(5):9-11

[27] Xiao Y-Y. Development strategy of Wugong Mountain scenic area in Pingxiang based on tourist satisfaction. Tourism Overview (Second Half of the Month). 2013;**20**:111-112

[28] Li L-P et al. Influence of tourism festivals on the mountain meadow: A case study of international Wugong Mountain tent festival. Biological Disaster Science. 2014;**37**(4):341-346

[29] Li F-X et al. Effects of grazing intensity on biomass and soil physical and chemical characteristics inalpine meadow in the source of three rivers. Pratacultural Science. 2015;**32**(1):11-18

[30] Huang X-X et al. Tolerance of alpine meadows to human trampling. Acta Prataculturae Sinica. 2014;**23**(2):333-339

[31] Li Z. Study on Species Diversity and Physical and Chemical Characteristics of Soil of Planted Forestin Degraded Red Soil Areas. Nanchang: Jiangxi Agricultural University; 2012

[32] Wang C-T et al. Relationship between plant communities, characters,

soil physical and chemical properties, and soil microbiology in alpine meadows. Acta Prataculturae Sinica. 2010;**19**(6):25-34

[33] Wu J et al. Effects of "habitat island" area on plant species diversity and its distribution pattern in Horqin sand land. Chinese Journal of Ecology. 2017;**36**(1):43-53

[34] Cheng X. Studies on Plant Community Characteristics and Spatial Distribution Attern in Subalpine Meadows of Wugong Mountain. Nanchang: Jiangxi Agricultural University; 2014

[35] Zhou G et al. Effects of changes in straw chemical properties and alkaline soils on bacterial communities engaged in straw decomposition at different temperatures. Scientific Reports. 2016;**6**:22186

[36] Ragot S-A et al. Total and active microbial communities and phoD as affected by phosphate depletion and pH in soil. Plantand Soil. 2016;**408**:15-30

[37] Tao Y et al. Leaf and soil stoichiometry of four herbs in the Gurbantunggut Desert, China. Chinese Journal of Applied Ecology. 2015;**26**(3):659-665

[38] Wei B-M et al. Physical degradation characteristics and mechanism of orchard soil in Weibei region. Journal of Plant Nutrition and Fertilizer. 2015;**21**(3):694-701

[39] Ma Y-J et al. Increase of soil nutrients and enzymatic activity by adding nitrogen and phosphorus to Larix principis-rupprechtii plantation. Journal of Plant Nutrition and Fertilizer. 2015;**21**(3):664-674

[40] Xu X-W et al. Effects of combined application of organic and chemical fertilizers on the yield of peanut, soil available nutrient and biological properties in the upland red soil in subtropical China. Acta Ecologica Sinica. 2014;**34**(18):5182-5190

[41] Wang E-W et al. Analysis on soil nutrients difference at different altitudinal in tobacco soil. Modern Agricultural Science and Technology. 2013;**21**:226-227

[42] Zhang L-L et al. Impact of grassland improvement measures on productivity of degraded grassland. Inner Mongolia Prataculture. 2012;**24**(3):37-46

[43] Zhang W-H et al. The feature analysis for grassland degradation and the restoration of natural vegetation in degraded grassland. Environment & Development. 2011;**23**(8):40-44

[44] Li Z-Z et al. Hyperspectral characteristics of main community types in Wugong mountain meadow. Pratacultural Science. 2016;**33**(8):1492-1501

[45] Deng B-L et al. Distribution patterns and relationships of organic matter, total nitrogen, and alkaline nitrogen in the grassland soil of Wugong Mountain. Jiangsu Agricultural Sciences. 2015;**43**(11):414-417

[46] Yuan Z-Y et al. Relations and distribution of soil organic carbon and Total carbon in different vegetation communities of grass zone in Wugong Mountain. Southwest China Journal of Agricultural Sciences. 2015;**28**(5):2166-2171

[47] Li Z et al. Effects of elevation and tourism disturbance on meadow soil infiltration on Wugong Mountain. Acta EcologicaSinica. 2018;**38**(2):635-645

[48] Deng B et al. Increases in soil CO_2 and N_2O emissions with warming depend on plant species in restored

alpine meadows of Wugong Mountain, China. Journalof Soilsand Sediments. 2016;**16**:777-784

[49] Yuan Y-D et al. Soil nitrogen distribution and its correlation with soil physical properties in different altitude in mountain meadow of Wugong mountain. Journal of Central South University of Forestry & Technology. 2016;**36**(10):108-113

[50] Bai J-L et al. Research progress of the restoration and reconstruction of degraded ecosystems. Journal of Zhejiang Forestry College. 2005;**22**(4):464-468

[51] Yang Y-G et al. Research on rapid turf establishing techniques. Pratacultural Science. 2006;**23**(1):93-96

[52] Ge Q-Z et al. Influence of restoration measures on plant community in alpine meadow. Pratacultural Science. 2012;**29**(10):1517-1520

[53] Yang Y. Characteristics and Quality Evaluation of Artificial Vegetation Recovery for Highway Slopes in Wanxi Dabie Mountain Areas. Beijing: Beijing Forestry University; 2016

[54] Li L-P et al. Research on the implementation of green Marketing in Wugong Mountain Scenic Area. Journal of Central South University of Forestry & Technology(Social Sciences). 2015;**9**(4):23-26

[55] Li Z et al. Growth effects and adaptabilities of different vegetation restoration measures of degraded meadow in Wugong Mountain. Journal of Central South University of Forestry & Technology. 2018;**38**(2):90-96

[56] Xiong X-Y et al. Study on protection and utilization strategy of Forest tourism resources in Henan Province. Heilongjiang Agricultural Sciences. 2021;**4**:119-124

[57] Hu M-W et al. Evaluation of the recreation environmental carrying capacity of mountain meadow: Taking Wugong Mountain scenic area as an example. Acta Agriculturae Universitis Jiangxiensis. 2016;**38**(6):1196-1204

[58] Hu M-W et al. Analysis on the characteristics of Tourists Mountain recreation to Wugong Mountain. Journal of Pingxiang College. 2014;**31**(5):28-34

[59] Huang P-F et al. The construction of greenways in mountain meadow scenic areas—Wugong Mountain scenic area as an example. Pratacultural Science. 2017;**34**(2):293-301

[60] Dong L-L et al. Tourist garbage treatment under festival activity background in mountain scenic areas: A case study of international tent festival of mountain Wugong. Biological Disaster Science. 2016;**39**(1):67-73

[61] Hu M-W et al. The culture production mechanism of the ancient trees in the context of tourism development. Forestry Economics. 2016;**3**:93-96

Chapter 7

Restoration of Uranium Tailings in the Mountains: A Perspective from the State of the Skin in Persons Living in the Vicinity

Alena Isupova

Abstract

In our work, for the first time, the problem of long-term exposure (more than 50 years) to low doses of radiation to the human body was touched upon. Comprehensive studies of the content of radionuclides in water, soil, meat of cattle, and small cattle, carried out within the framework of the ISTC; prove the entry of radionuclides into the human body through food chains. For the period from 2003 to 2023, studies were conducted on the skin of people living in various regions of Kyrgyzstan. A total of 3650 people were examined. The main group (600 people) is represented by people living near uranium tailings. Analysis of the features of the course and manifestation of dermatoses showed the features of the course of skin pathology. The mean age of stigma manifestation was 32.7 ± 7.5 years compared to the control group 45.2 ± 10.3 years.

Keywords: radionuclides, internal exposure, skin, stigmas of age-related skin changes, body aging, skin aging, uranium tailings, skin microbiome, skin diseases, immunity, pyoderma, diagnosis of skin neoplasms, epiluminescence microscopy, dermatoscopy, uranium tailings, 25(oh)D, skin microbiota

1. Introduction

Problems of environmental safety and public health in technogenic zones located in the mountains are of great scientific and practical importance.

Radiation safety is the main issue of environmental safety and public health in Kyrgyzstan. The legacy of past activities in the mining and processing of uranium remains relevant in connection with the spread of industrial waste throughout the territory of the republic and border countries [1]. Depleted uranium (DU) obtained as a result of processing natural uranium makes a large contribution to environmental pollution [2].

On the territory of Kyrgyzstan (199,951 km^2) there are 49 tailings and 80 rock dumps, where 70 million m^3 of uranium production waste is buried. In industrial tailing waste, in addition to uranium, there are radioactive elements in high concentrations, such as radium - 226, thorium - 230, and radon - 222, as well as elements that were used as reagents in the processing of ore: Ca, Si, Pb, Cr, Mn, V, Ni [3]. Most of the tailings are located in the mountain ecosystem, above settlements, near river mouths, or seasonal water flows. The lack of control over the movement of industrial waste has a negative impact on individual components of the mountain ecosystem, including plants, animals, and humans [4, 5]. Climate change is possible under the influence of anthropogenic factors; projected global warming can increase the number and severity of natural disasters in the mountain ecosystem [6]. In emergency situations, there is a danger of transferring industrial waste to the territory of neighboring states. The peculiarities of the spread of industrial waste are largely related to the high mountainous relief of the republic, seismic activity, seasonal melting of glaciers and mudflows, and wind erosion of soils. An analysis of the situation shows that more than 26,000 people are in the zone of ecological catastrophe in Kyrgyzstan, about 2 million people in Uzbekistan, 900,000 people in Kazakhstan, and 700,000 people in Tajikistan [7].

According to studies, drinking water from artesian wells was characterized by an increased content of U (up to 10 μg/l) and some microelements (for example, As, Se, Cr, V, and F), especially arsenic. In the drainage waters of the tailings, uranium concentrations exceeded those recommended by WHO by more than 200 times, and a large scatter was observed in the change in the (234)U/(238)U isotope ratios. Another study showed that the ingestion dose of radioactive radiation for members of the critical group is from 10 to 30 mSv, and in an emergency scenario - 45 and 77 mSv for an adult and a child, respectively [8].

The concentrations of radionuclides and chemical elements in the water of Lake Issyk-Kul were generally low but unexpectedly high for arsenic. Uranium (U), As, and Ni were mostly present in all samples as low molecular weight particles (<10 kDa), indicating that these elements are mobile and potentially bioavailable. Soils at mining sites are enriched with uranium, arsenic, and other elements. In Kaji-Sai, hotspots with increased levels of radioactivity were easily found, where accumulations of radioactive particles were observed. The presence of particles carrying a significant amount of radioactivity and toxic microelements can be dangerous during strong wind events (wind erosion) [9]. According to Salbu B., Burkitbaev M. et al. field expeditions to the Kurdai mine (border of Kazakhstan, 23 km from Bishkek) showed a high content of man-made waste in water and soil. Uranium, As, Mo, and Ni were predominantly present in waters in the form of mobile low molecular weight forms, while a significant part of Cr, Mn, and Fe was associated with colloids and particles. It was noted that, due to the oxidation of ferrous iron in artesian groundwater, upon contact with air, iron serves as a sink of other elements, peak concentrations of uranium, radium, arsenic, and manganese isotopes are observed, and based on successive extractions, a significant part of U, Pb, and Cd can be considered mobile [10]. A similar situation with the impact of radionuclides and metals on the environment is observed at the former U-Taboshar and Digmay test sites in the neighboring republic [11].

Insufficient security associated with free access to uranium tailings can pose a serious threat to the spread and use of radioactive components of former enterprises [12]. In many technogenic zones, cattle grazing was noted near burial grounds (Figure). Local residents visit open adits (**Figure 3**) and extract building materials and coal for the needs of the family.

Today in Central Asia there is a problem of limited data on water quality [13], taking into account climate change and the man-made introduction of humans into the mountain ecosystem. Climate change entails a change in the quantity and quality of water resources in the Central Asian republics [14]. On a global scale, the flow of rivers originating in the mountains increases in response to climate warming and the depletion of glaciers [15]. Large uncertainties in meteorological datasets remain a major obstacle to understanding the processes that cause changes at the interface between climate, glaciers, runoff, and mountain ecosystems as a whole [16].

A study by Hao et al. [17] proposes a new analytical framework for assessing the security of water, energy, and food (WEF) and the factors influencing them in five countries of Central Asia: Kazakhstan, Kyrgyzstan, Turkmenistan, and Uzbekistan; Tajikistan. At the same time, four dimensions of security indicators are proposed: availability, self-sufficiency, productivity, and accessibility [17]. It is hoped that radiobiology and radioecology will be included in the study of factors affecting the safety of WEF. There is no better opportunity to study in detail the mechanisms of action, biological effects, kinetics in the body, and ways of excretion of radionuclides, as their wide application in medicine. Beta-particle emitters are used, however, a large radius of action (≥1 mm) can cause nonspecific irradiation of healthy cells [18]. Radionuclides emitting predominantly alpha particles have a narrowed exposure range to 50–100 μm [19]. Although it is known that the emitters of alpha particles decay along the decay chain, which leads to the appearance of several daughter particles, causing additional biological effects [20]. Auger electrons are very low-energy electrons emitted by radionuclides that decay by electron capture. This energy is transferred over distances ranging from nanometers to micrometers, resulting in a highly linear energy transfer capable of causing lethal cell damage. These high-energy deposits produce local dense ionizations that are highly destructive when the radionuclide is incorporated into nuclear DNA, the cell nucleus, or the cell membrane [21].

The International Atomic Energy Agency (IAEA) plays an important role in controlling the quality and mechanisms of action of radionuclides on a living organism. The Agency is pursuing the issue of “Preclinical testing of radiopharmaceuticals” to minimize the risk of unknown and undesirable effects [22]. Such work could make a great contribution to the problem of the health of people living in the zone of technogenic pollution for a long time, taking into account the cross-understanding of research topics and the symbiosis of sciences [23]. It is known that radionuclides penetrate into the internal environment of the human body with water, food, and also by inhalation of aerosols [24]. The clinical effects of exposure to DU, like other heavy metals, are manifested by nephrotoxicity [25], neurological symptoms, immunotoxicity, embryotoxicity [26], and hepatotoxicity [27]. Epidemiological [28] and experimental studies confirm an increased risk of neoplasms [29].

Assessment of radiological doses due to endogenous intake of radionuclides is calculated using biokinetic and dosimetric models [30]. IMBA is a set of software modules that implement current ICRP (International Commission on Radiation Protection) biokinetic and dosimetric models for intake and dose assessment. When calculating, IMBA applies the truncation rule for progeny in radioactive series [31], which limits the results of the study. There are recommendations of the ICRP in case of nuclear accidents, taking into account the scale and necessary radiological protection and, in the long term, the development of measures to restore the living conditions, working conditions, and quality of life of the affected communities. To achieve this goal, the ICRP emphasizes the critical importance of bringing all stakeholders together [32]. In our opinion, following this example, monitoring the situation at uranium tailings dumps and the safety of the population in technogenic zones should also become a priority for stakeholders and the state.

It is known that the incorporation of radionuclides into the internal environment of the body will not cause acute radiation sickness, but it can cause long-term internal exposure, leading to health consequences after several years or even decades. In this variant of exposure, the harmful biological effects of radiation do not have a dose threshold of occurrence and are stochastic. With increasing doses, it is not the severity of these effects that increases, but the likelihood (risk) of their occurrence [33]. The health risk can be quantified by the committed effective dose, which is defined as the cumulative effective dose from internal exposure 50 years (for adults) after the incorporation of the radionuclide [34]. In addition, uranium is a heavy metal (19.1 g/cm^3) and therefore may have chemically induced toxic effects [34]. The mechanisms of heavy metal toxicity are quite complex [34]. The toxicity and absorption of uranium (concentration in the target tissue) depends on its soluble and insoluble forms [35]. The mechanisms underlying the toxicological effects of uranium on the human body are being studied [36].

Thus, active technogenic pollution of the environment as a result of the mining and processing of uranium, as well as the disposal of uranium waste, significantly disrupts the existing balance of the abiotic part of the mountain ecosystem and the biocenosis, which consists of the exchange of matter and energy. Our scientific research is devoted to studying the influence of "low doses of radiation", "toxicological profile" of radionuclides on the state of the integumentary system, immunological status, and skin microbiota of communities living for a long time (more than 50 years) near uranium tailings in a mountain ecosystem.

2. Research results

2.1 Territorial living conditions of a cohort of Kyrgyz in a mountain ecosystem

The research was carried out within the framework of the scientific project KR766 of the International Science and Technology Center (ISTC). To record the radiation level, the participants of the ISTC KR766 project expedition used a domestic radiometer SRP-68-01 type RPTU-01 with a BTGI-01 sensor, which allows recording gamma radiation at a discrimination energy level of 20 ± 5 keV and with a measurement range of exposure dose rate up to 3000 μR/h [5].

Scientific research covered the territories of the geochemical provinces of Min-Kush, Mailu-Suu, and Kadzhi-Sai of the Republic of Kyrgyzstan. The country's territory is located within two mountain ecosystems. Its northeastern part lies within the Tien Shan, and the southwestern part lies within the Pamir-Alai. Kyrgyzstan is 93% mountainous and mostly lies on land, located at an altitude of 1000–7400 m. The average altitude above sea level is 2750 m. More than 40 percent of the country's territory is at an altitude of more than 3000 m, and three-quarters of this is covered by permanent snow or glaciers, with 600 glaciers covering an area of 6578 km^2 [37].

The geochemical province Min-Kush ("thousand birds") is located in the spurs of the Moldo-Too ridge in the geographical zone of the Central Tien Shan, at an altitude of 2200 to 2500 m above sea level (**Figure 1**). The figure shows a mountain ecosystem, where a settlement is located in a gorge along the river, and uranium tailings are located in the upper reaches of the mountains. Background radiation levels in the urban area of Min-Kush compared to ecologically "clean" areas of the republic (18–22 μR/h) are increased: three times higher (60–61 μR/h). Thus, at the gates of the Tuyuk-Suu tailings dump, the background radiation level was 60–61 μR/h, on

Figure 1.
The geochemical province of Min-Kush.

the territory of the Rudnik hotel - 60 to 61 μR/h, in the water flowing from the adit. No. 8–61-61.5 μR/h.

Geochemical province Kadzhi-Sai (**Figure 2**) on the southern shore of Lake Issyk-Kul, at an altitude of 1750–1978.9 m above sea level. **Figure 2** shows a populated area with uranium tailings and rock dumps located above it.

On the right of the photo, you can see the entrance part of the adit. It is important to note that the adits remained unguarded for a long time. The entrance to the adits is open (**Figure 3**) and accessible to animals, residents, including children.

Figure 2.
Geochemical province of Kaji-Sai located near the mountain lake Issyk-Kul.

Figure 3.
The entrance to the mine adits is open.

Figure 4.
The rock dumps and uranium tailings.

Figure 5.
Mudflow passing through the territory of the uranium tailing dump toward residential areas and the lake.

Figure 4 shows rock dumps, and on the right is a fragment of wastewater from a uranium tailings pond. Our expedition vehicle is visible in the background. The radiation level here is from 600 to 1000 μR/h.

Figure 5 shows a large wastewater flow from a uranium waste disposal site toward a populated area and mountain lake in this mountain ecosystem.

Mountain rivers and mudflows violate the integrity of uranium tailings and thereby spread man-made pollutants throughout residential areas, fields, gardens, and pastures, and these waters end up in Lake Issyk-Kul.

In 1948, there was a uranium ore processing plant here. Uranium oxide was extracted from the ash of brown uranium-containing coals of the Sogutinsky deposit after their combustion (uranium oxide was extracted by leaching from the ash). In the Kadzhi-Say geochemical province, the radiation level is relatively low and ranges from 18 to 40 μR/h, but the uranium content in the surface waters of the Kadzhi-Say tailings pond, as well as on the southern shore of Lake. Issyk-Kul, adjacent to the province, is 1–2 orders of magnitude higher than on the opposite shore of the lake. If in the water of the northern shore of Lake Issyk-Kul the uranium content is about $0.3–4.94 \times 10–^{5}$ g/l, then in the region of the Kadzhi-Sai province this figure is 2–60 times higher. A total of $11.7 \times 10–^{5}$ g/l of uranium was found in tap water from the well of the Khan-Saray boarding house, compared to tap water in the village of Komsomolsky, Chui region, where the radioecological laboratory is located, containing uranium $1.3 \times 10–^{5}$ g/l.

The geochemical province Mailu-Suu (**Figure 6**) is located in the Jalal-Abat region on the slopes of the Fergana Range at an altitude of 1100–1200 m above sea level. This is the largest uranium province, in which a uranium processing plant operated from 1946 to 1967, and its waste was stored along the right and left banks of the Mailu-Suu River. After the closure of the mining plant, 23 uranium tailings and 18 dumps of substandard uranium ores remained in the region. 1.9 million m^3 of spent uranium ore are buried in this area.

The radiation background of gamma radiation in the geochemical province of Mailu-Suu ranged from 16 to 600 μR/h, and according to the regional department of the Ministry of Emergency Situations of the Kyrgyz Republic, in some areas, it reached up to 1000 μR/h, i.e. exceeded the radiation levels existing in clean zones by 2–4 times, and in anomalous points by 20–50 times.

Figure 6.
Former uranium mining site of Mailuu-Suu [38].

Figure 7.
Pastures and hayfields in the mountain ecosystem of Kyrgyzstan.

2.2 Analysis of food products obtained from livestock farming in a mountain ecosystem near uranium tailings

The expedition members examined the organs of large and small livestock in the three main uranium provinces of the republic. Uranium content in the organs of Min-Kush cows: 0.001–5.54 mg/kg wet weight. In the city of Mailu-Suu, it is 0.002–0.212 mg/kg and in the city of Kadzhi-Sai - 0.002-0.537 mg/kg wet weight.

The uranium content in the organs of small ruminants (lambs) in the city of Min-Kush ranges from 0.005 to 2.44 mg/kg, in the city of Mailu-Suu - 0.03-0.107 mg/kg, in the city of Kadzhi-Sai - 0.001–0.048 mg/kg. The uranium content in integumentary tissues (skin, horns, hooves) ranges from 0.003 to 0.18 mg/kg wet weight.

Figure 7 shows the grazing of large and small livestock on natural pastures, part of the territory of which includes unprotected tailings ponds and mountain dumps. On the right, you can see grazing animals among the rock dumps. In the center of this figure, the water flow passing through the rock dumps is clearly visible.

Thus, the material we obtained convincingly proves that radionuclides enter the body of animals and humans through food chains.

2.3 General characteristics of uranium burial sites in the mountain ecosystem of Kyrgyzstan

Thus, waste from production and industrial equipment with a total volume of about 400,000 m^3 is buried on the territory of the republic. After the cessation of the activities of mining enterprises for the extraction and processing of uranium, 35 of the 49 tailings dumps that formed turned out to be ownerless. Of the 80 dumps of substandard uranium ores, 25 are abandoned.

Mountain rivers and mudflows violate the integrity of uranium tailings and thereby spread man-made pollutants throughout residential areas, fields, gardens, pastures, and these waters end up in rivers and lakes, which causes enormous damage to the mountain ecosystem as a whole. Foothills and mountain settlements suffer from the presence of uranium tailings and rock dumps.

2.4 Methods for studying communities living near uranium tailings in a mountain ecosystem

During the period from 2003 to 2023, 3650 people living in various regions of Kyrgyzstan were examined. The risk group is people living near uranium tailings in the mountains; the comparison group – people living in radionuclide-free areas.

The skin was examined using standard methods. Initially, the condition of externally unchanged areas of the skin was assessed, and then the lesions were described. Skin color and the presence of de- and hyper-pigmentation were determined, indicating the place where these changes were most pronounced; elasticity, turgor, and hydration (normal, dry, moist skin). If rashes were present, their nature (inflammatory or non-inflammatory), symmetry, extent of the process, localization of the rash, and damage to open or closed areas of the skin were assessed.

To analyze clinical and epidemiological studies, a specially designed card was filled out based on information obtained from the study of outpatient records, clinical and instrumental methods for studying the skin of selected groups. During the interview, a questionnaire consisting of 54 questions was filled out. Clinical assessment of pigmented skin lesions was carried out according to the ABCD rule (R. Friedman) [39].

Epiluminescence microscopy of skin neoplasms was performed using a Heine Delta 20 dermatoscope (K-256.27.376, Heine Optotechnik, Germany), an emersion medium (Dermatoscopy Oil, K-00.34.005, Heine Optotechnik, Germany), a recording device - a Nicon 5300 digital camera (Japan); photo adapter designed for contact with a HEINE DELTA 20/Nikon Coolpix writing device (K-00.34.235, Heine Optotechnik, Germany). A dermoscopic assessment of melanocytic pigmented skin lesions was performed using “pattern analysis” [40]. Panoramic skin scans were taken under standard conditions.

2.4.1 Immunological studies

Immunological studies included the study of T- and B-links of immunity, the phagocytic activity of neutrophils, and the system of mononuclear phagocytes. The content of T- and B-lymphocytes and subpopulations of T-cells in the blood was determined by the method of indirect surface immunofluorescence with monoclonal antibodies of the IKO series, a “shortened” panel was used to identify CD markers: CD3+ (T-lymphocytes); CD4+ (helper T-lymphocytes); CD8+ (cytotoxic lymphocytes); CD20+ (B-lymphocytes). The study of mononuclear phagocytes was carried out using the following methods: (1) Determination of the phagocytic activity of monocytes in a test with monodisperse latex particles with a diameter of 1.0–2.0 μm; (2) Nitroblue tetrazolium test in monocytes (NBT-test); (3) Adhesion of monocytes; (4) Spreading of monocytes; (5) Determination of the total luminescence index (SIL) of monocyte lysosomes; (6) In some cases, lipid inclusions were detected in the cytoplasm of monocytes using a luminescent probe 3-MBA (3-methoxybenzanthrone). In parallel, the phagocytic activity of neutrophils and the intensity of their reduction of nitroblue tetrazolium were studied. (7) The calculation of the relative and absolute content of monocytes in the peripheral blood was carried out according to the generally accepted method.

2.4.2 Study of skin microflora

To study the quantitative and qualitative composition of the skin microflora, 67 skin samples were studied on Korostelev’s medium, and 53 samples were taken by washing. The respondents were conditionally divided into three groups: (1) residents of the clean zone (28 people); (2) residents of the technogenic zone (30 people); (3) ormer miners of a uranium mine (28 people).

When studying the ecology of the skin of the examined groups, appropriate methods of material selection were used. The isolation of microorganisms from their natural habitat - human skin - was carried out by washing and imprinting, which made it possible to count viable microcolonies or clusters of bacteria located on the surface of the skin (a quantitative method for studying surface microflora). This method refers to direct methods of collecting material and is convenient and easy to perform. When using the direct method, slides with Korostelev's medium (per 100 ml of meat peptone agar, 0.5 ml of a 1.5% alcohol solution of bromothymol blue) were tightly placed for 2 seconds on the surface of the skin of the shoulder, placed in petri dishes and incubated in a thermostat at a temperature of 37°C 24 hours. One cup of medium not in contact with the skin was left in each dish to control airborne flora that may enter when the lid is opened (usually airborne flora does not grow on this medium for 24 hours at 37°C). To differentiate bacteria by biological properties, 1% mannitol was added to Korostelev's medium with bromothymol blue solution.

Microbes that did not decompose mannitol formed colonies colored in the color of the medium—dark green (**Figure 8**), and those capable of fermenting it— bright yellow. (Fermentation of mannitol is one of the signs of microbial virulence.)

Figure 8 shows preparations with colonies of skin microflora of examined individuals living near uranium tailings dumps and in the "clean zone". The appearance of plates of Korostelev's medium with colonies of skin microflora (yellow colonies - mannitol-positive, green colonies - non-fermenting mannitol).

A washing method in which dispersion of aggregates is achieved and individual cell counting is subsequently possible, which makes it possible to study the deep microflora of human skin. In this method, a rod was placed in 1 ml of 0.85% saline solution, and a stencil with an area of 1 cm^2 was applied to a sterile cotton swab. Further seeding of the material was carried out on artificial nutrient media. To study microbes from the skin of subjects, cultures were obtained in the form of homogeneous populations (pure cultures) by dispersing mixtures onto other media of varying consistency, both to isolate the colony and to preserve the culture or to study its enzymatic properties. Isolated microbial cultures were identified by studying the morphological characteristics of bacterial colonies and the nature of growth on solid and liquid nutrient media. Biochemical features include the ability of microorganisms to ferment glucose, lactose, sucrose, maltose, and hexahydric alcohol mannitol. The pure culture was sown in a loop onto the "variegated row" media. The results of the work to identify the isolated culture based on morphological and physiological characteristics were presented in a standard form.

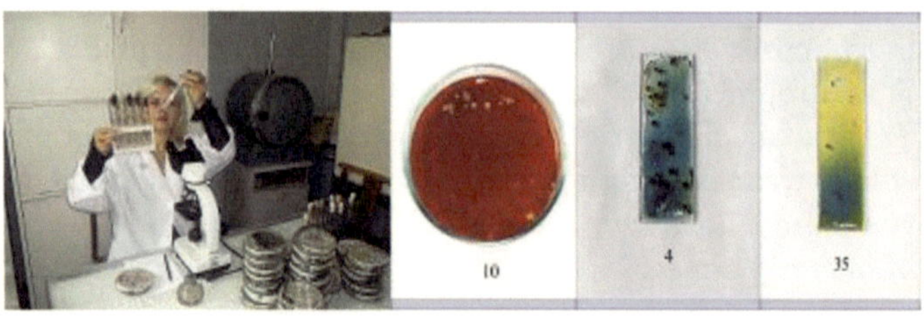

Figure 8.
Microbiological studies were carried out in the laboratory of the Department of Microbiology of KSMA.

2.4.3 Serum 25(OH)D level testing

The level of 25(OH)D in nmol/l was studied on a Cobos e-801 analyzer (Roshe, Italy) using the enzyme immunoassay method. Coefficients of variation for 25(OH) D ranged from 4.0% to 6.1%, indicating the validity and reliability of the measurements. Internal quality control is confirmed by the program (TIQCON) of Roshe, Italy.

2.4.4 Data collection and use of variables

All subjects signed informed consent to participate in the study. All ethical standards and principles of confidentiality were maintained during the study in accordance with the principles of the World Medical Association Declaration of Helsinki, as amended in 2000, "Ethical Principles for Medical Research Involving Human Subjects."

2.4.5 Statistical analysis

Statistical analysis was performed using PASW Statistics 21.0 (SPSS Inc., IBM, Chicago, USA). Descriptive statistics are presented as mean, standard deviation (SD), and standard error (SE). The statistical significance of the variables was determined using Pearson's chi-squared test. All tests are two-tailed, and $p < 0.05$ was considered statistically significant. The contingency coefficient phi Cramer̍ s V was used to assess the relationship of nominal data, where the value varies between 0 and 1, with "0" indicating no relationship between the row and column variables, and a value close to 1 indicating a high degree of relationship between these variables. To assess the effect of exposure to a risk factor, attributable risk (AtR), risk ratio (RR), and potential harm index (NNT) were calculated for each variable. For all variables, a 95% confidence interval (95% Cl) was calculated. Statistical significance was determined using the Cochran and Mentel-Henzel test, where $p < 0.05$ was considered statistically significant. After the Bonferroni correction (when >2 groups were set), the significance level would be $p < 0.01$.

2.5 Health indicators

2.5.1 Incidence analysis

In children of the geochemical uranium provinces of the cities of Kaji-Sai, the cities of Min-Kush, the cities of Mailu-Suu up to 14 years old (650 people), based on the diagnoses made in outpatient cards and annual reports according to the data for formation No. 12, it can be seen that in the first place are respiratory diseases - 58% (of which 48% are acute respiratory infections); in second place - diseases of the oral cavity and teeth, then blood, skin, endocrine system. In the adult population (800 people), diseases of the digestive system (52%), and respiratory organs (47%) are in the first place, then, in combination with the previous nosological groups, diseases of the cardiovascular and genitourinary systems. Comorbidity in the adult population was noted in 38% of cases. Complaints. In children, in the first place are frequent colds, dryness, peeling and redness of the skin, itching, aggravated after water procedures, and the presence of pigmentation. In adults, asthenic conditions, especially in the afternoon, pain in muscles and joints for no apparent reason, frequent headaches, and aching in nature. On the part of the skin, there are complaints of early graying of hair, in some from the age of 13–14 years, severe dryness of the skin, itching, aggravated after the use of soap and detergents, poor wound healing, the presence of neoplasms.

2.5.2 Skin condition of residents of technogenic zones and former miners of a uranium enterprise

In children of younger preschool age, viral diseases of the skin and mucous membranes, superficial pyoderma, and nonspecific hand dermatitis were more often recorded; in middle and older preschool age in 36% of cases, a combination of signs was revealed: xerosis and peeling of the skin; itching of the skin of varying severity, aggravated by sweating; erythematous-squamous rashes on the skin of the elbows and in the popliteal fossae; in school-age children, the following combination of symptoms was revealed in 27% of cases: erythematous-squamous rashes with lichenification, severe dryness, peeling and itching of the skin, forehead peteriasis, cheilitis, nonspecific hand dermatitis in spring and summer, nonspecific dermatitis of the feet in winter. In 18.6% of cases, a bacterial infection joins the violation of the epidermal barrier. Many children have a reaction from the mucous membranes of the nose, larynx, and cornea of the eye to the flowering of plants, especially wormwood. This is manifested by redness of the eyes, tearing, sneezing, a sensation of scratch in the throat, and in some cases a dry cough.

In the adult population, a bacterial infection of the skin comes to the fore, dyschromia is in second place, then benign neoplasms, and photodermatosis, as one of the common causes of skin pigmentation disorders in the examined individuals; in the future - atrophic changes in the skin and its appendages, fungal skin lesions, vascular atrophic poikiloderma.

A feature of bacterial infections of the skin in residents of technogenic zones and miners is the torpidity of the course with a predominance of the hemorrhagic component and mild infiltration. Chronic ulcerative pyoderma is present in 3.1% of cases; miners in 43% of cases with deep pyoderma did not have a pronounced pain syndrome. In 10% of patients, the course of the disease exceeded 15 years, relapses were observed 3–5 times a year. Among fungal infections, infiltrative-purulent forms of the beard area (in men) and pubis, onychomycosis, candidiasis, and multi-colored lichen prevail. In some patients (15.7%), the course of multi-colored lichen was of a generalized nature. Usually, multi-colored lichen occurs against the background of increased sweating, which was not observed in the examined individuals. When examining the integumentary system of a person, increased dryness of the skin (xerosis) was noted, the hair of 48% of the inhabitants of technogenic zones and 78.2% of the miners were dry, dystrophic, and 8.3% had diffuse alopecia. Patients. In 39% of cases, various types of nail dystrophy were noted. At a young age, koilonychia, onychomadesis, micronychia are more often observed; in 3%, canal-like dystrophy of the nail plates of the thumbs is detected. Onycholysis, onychoshisis, and onychogryphosis are more often observed in people over 30 years old in 4.1% of residents of technogenic zones and 13.9% of former miners. Thinning of the skin and the presence of a pronounced vascular pattern of the skin were noted. In 9% of cases, in combination with skin dyschromia. In Bishkek residents, such skin changes are sporadic. It is interesting to note that against the background of dry and atrophic skin, hyperplastic processes were observed in the form of warts, papillomas, and seborrheic keratomas.

2.5.3 Bacterial contamination of the skin in residents of uranium geochemical provinces and former miners of a uranium enterprise

Violation of the epidermal barrier under conditions of environmental instability leads to quantitative and qualitative changes in the commensal flora of the skin (**Table 1**). Violation of skin microbiocinosis entails a more severe course of skin diseases due to contamination with pathogenic flora (**Figure 9**).

Continuous growth was observed in 23.1 ± 1.9 and 35.7 ± 2.5 with a predominance of mannitol-positive flora in 53.8 ± 3.4 and 57.1 ± 2.2 in the demonstration group and in the group of miners, respectively, which indicates a violation of skin microbiocinosis (**Table 1**).

Healthy skin is characterized by a high population of commensal flora and low species diversity (**Figure 9**); a shift in species diversity of the skin flora is one of the signs of a violation of microbiocenosis.

Groups by number colonies on medium plates Korosteleva	1 Control group living in a clean area (n = 28)	2 Persons living in conditions of technogenic pollution (n = 30)	3 Miners (n = 28)
0–50, %	88.9 ± 2.3	57.6 ± 3.1	28.6 ± 1.9
51–100, %	11.1 ± 1.4	19.2 ± 2.2	21.4 ± 2.0
Over 100, %	0	23.1 ± 1.9	35.7 ± 2.5
Solid growth, % of them mannitol-positive bacteria (in % of the total number)	0 14.8 ± 2.9	0 53.8 ± 3.4	14.2 57.1 ± 2.2
$p < 0.01$.			

Table 1.
Level of bacterial contamination of the skin in residents of uranium geochemical provinces and former miners of a uranium enterprise (M ± m).

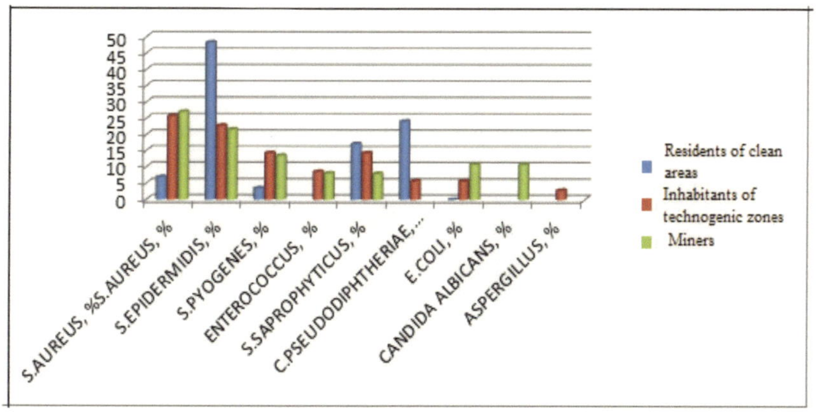

Figure 9.
Species composition of skin microflora in residents of the uranium geochemical province and former miners of the uranium enterprise.

2.5.4 Immune system

Normally, the ratio of T- and B-lymphocytes is 70–80%/10–15% of blood lymphocytes. In residents of the clean zone, this ratio changes from normal values toward a decrease in T-lymphocytes. Similar changes are occurring with the residents of the geochemical province. It is known that the phenotype of the subpopulation of T-helper cells (helpers) is represented by antigens CD3, CD4, and T-cell receptor antigens HLA class II. These lymphocytes recognize foreign antigens processed by macrophages, activate the mechanisms of activation of B lymphocytes, and produce antibodies (Th2). At the same time, the population of CD3 cells decreases, which leads to a decrease in the activity of T-cell immunity (**Figure 10**).

A simultaneous increase in the number of cytotoxic blood lymphocytes (CD8) leads to inhibition of the activation function of T- and B-lymphocytes and their reactions. Our observations show a tendency toward a decrease in the functional activity of B-lymphocytes and a decrease in the CEC in the studied groups (**Figure 10**). The population of B lymphocytes is the precursor of plasma cells that form protective antibodies of an immunoglobulin nature. Once activated by an antigen, B lymphocytes form memory cells, which allows the body to quickly synthesize large quantities of antibodies when repeated immunization with the same type of antigen.

A distinctive feature of the decrease in humoral immunity in residents of geochemical provinces in the mountain ecosystem and former miners was a characteristic decrease in the concentration of serum IgM and IgG, apparently associated with the depletion of the function of B-lymphocytes at the final stage of their differentiation stage, i.e. plasma cells (**Figure 11**). Circulating immune complexes and immunoglobulins in residents of a uranium geochemical province in a mountain ecosystem and former miners of a uranium enterprise in comparison with the control group.

The indicator of the nitroblue tetrazolium test (**Table 2**) decreases, which reflects the state of the bactericidal peroxidase systems of the cell and correlates with the formation of superoxide radicals, which is distorted by the relative lack of O2 in the inhaled air.

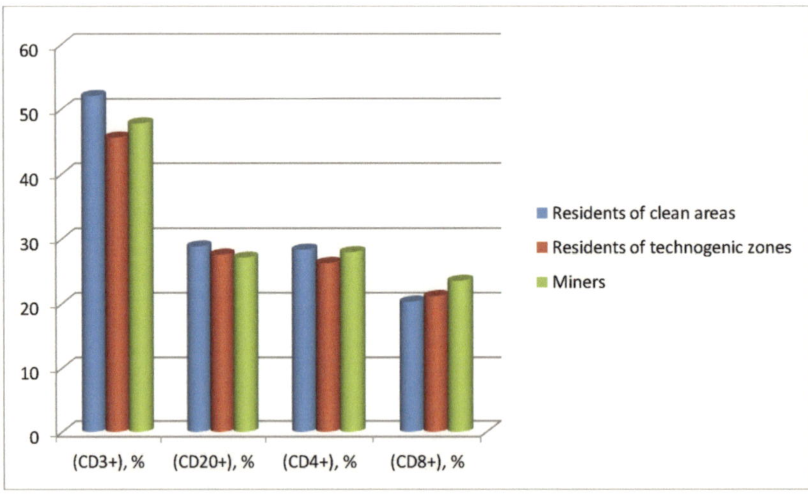

Figure 10.
Indicators of cellular immunity in residents of a uranium geochemical province located in a mountain ecosystem and former miners of a uranium enterprise in comparison with the control group.

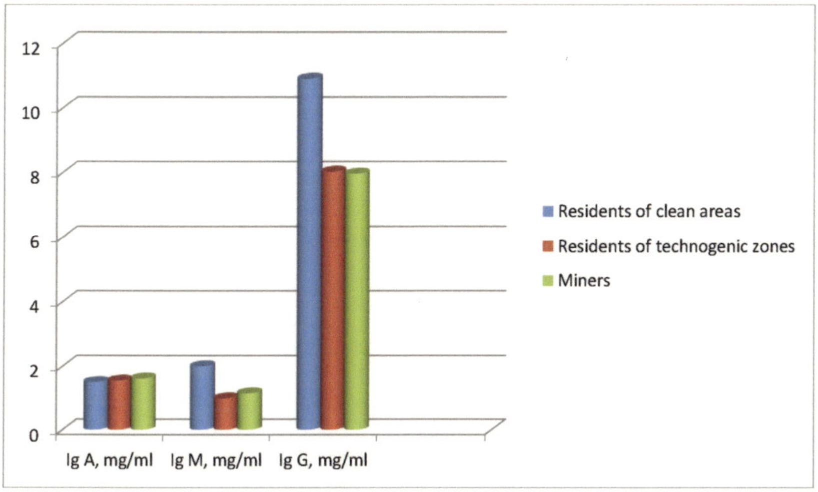

Figure 11.
Circulating immunoglobulins (Ig) in residents of a uranium geochemical province located in a mountain ecosystem and former miners of a uranium enterprise in comparison with the control group.

The lysosomal apparatus of mononuclear phagocytes is closely related to the processes of catabolism, the work of which is enhanced by radiation (**Table 2**). At the same time, we took into account that the content of lysosomes in the cytoplasm of monocytes is a reflection of oxygen-independent mechanisms for the destruction

Indicators	**1 Control group living in a clean area (n = 28)**		**2 Persons living in conditions of technogenic pollution (n = 30)**		**3 Miners (n = 28)**	
	Monocytes	**Neutrophils**	**Monocytes**	**Neutrophils**	**Monocytes**	**Neutrophils**
Phagocytic index, %	59.1 ± 4.2	70.3 ± 2.9	52.09 ± 1.53	62.3 ± 1.3	51.21 ± 2.27	64.75 ± 0.8
Phagocytic number	2.2 ± 0.9	3.3 ± 0.7	1.89 ± 0.09	2.31 ± 0.07	1.83 ± 0.19	2.25 ± 0.17
Integral phagocytic index	1.3 ± 0.01	2.3 ± 0.02	0.99 ± 0.06	1.45 ± 0.06	0.92 ± 0.11	1.54 ± 0.11
Nitroblue tetrazolium test, %	85.3 ± 0.3	90.1 ± 0.2	76.09 ± 1.36	81.5 ± 0.91	77 ± 2.09	85.75 ± 1.1
Average cytochemical coefficient	1.3 ± 0.04	1.7 ± 0.01	1.0 ± 0.05	1.46 ± 0.07	1.06 ± 0.01	1.41 ± 0.09
Adhesion, %	37.4 ± 2.7		20.61 ± 1.2		21.53 ± 1.46	
Spreading, %	29.2 ± 3.1		14.62 ± 0.81		14.1 ± 0.83	
Total luminescence index of lysosomes	329.8 ± 162		188.55 ± 8.5		202.08 ± 8.7	

$p < 0.01$.

Table 2.
Functional indicators of blood monocytes and neutrophils in residents of the uranium geochemical province and former miners of a uranium enterprise (M ± m).

of microorganisms, which also indicates a deterioration in the functional activity of monocytes. Thus, among residents of technogenic zones and former miners of a uranium plant, both mechanisms of bactericidal activity—oxygen-dependent and oxygen-independent—function at an extremely low level, not completely protecting the body from infection. Both indicators—"adhesion" and "spreading" are significantly reduced, which once again confirms the low functional activity of monocytes (**Table 2**).

We have established a clear connection between skin barrier disorders and skin microbiome dysbiosis with immune dysregulation and high sensitization to environmental allergens, which is manifested by a variety of skin symptoms characteristic of residents of technogenic zones.

2.5.5 Skin tumors in the risk group

Epidermal tumors - 58.2% (95% CI 56.5–59.9); melanocytic formations - 4.51% (95% CI 3.8–5.24); non-tumor pigmentation - 10.7% (95% CI 9.6–11.7); vascular tumors 17.7% (95% CI 15.8–18.5); fibrous formations 5.2% (95% CI 4.4–6.0); from nervous, muscular and adipose tissue 3.3% (95% CI 2.6–3.9). Melanoma and non-melanoma cancer 1.5% (95% CI 1.06–1.93) [41].

Calculation of Pearson's chi-square coefficient showed the existence of a relationship between the risk factor and the frequency of occurrence of the main stigmas of geranthogenesis in people living near uranium tailings. The calculation of the contingency coefficient phi Crameŕ s V showed the strongest relationship with the risk factor for seborrheic keratoses - 0.627. For other statistically significant variables, the association strength is moderate (from 0.3 to 0.6). This gives grounds to assert that these signs can serve as criteria for assessing the influence of a risk factor on age-related skin changes. The NNH calculation indicates that 1 out of 3 residents of the risk group may develop early age-related stigmas [42].

2.5.6 Study of serum 25(OH)D level using enzyme immunoassay

The average level of 25(OH)D in the blood serum of the examined was 38.7 ± 1.0 nmol/l (95% CI 36.6–40.7 nmol/l), and there were no statistically significant gender differences (p = 0.066). Seasonal fluctuations were established - from 30.2 ± 1.9 nmol/l (95% CI 26.3–34.1 nmol/l) in winter to 46.6 ± 1.9 nmol/l (95% CI 42.8–50.3 nmol/l) in autumn (p = 0.0001). The level of 25(OH)D less than 50 nmol/l was detected in 71.3 ± 2.6% (95% CI 66.1–76.5%) of cases, 50–75 nmol/l — in 24.6 ± 2, 5% (95% CI 19.7–24.5%) of cases, more than 75 nmol/l in 4.1 ± 1.2% (95% CI 1.8–6.4%) of cases [43].

2.5.7 Discussion

Communities living for a long time (more than 50 years) in conditions of technogenic pollution are characterized by daily endogenous intake of increased amounts of toxic and radioactive substances (radionuclides) through the food tract, transdermally and aerogenously. Arguing about the pathogenetic mechanisms of the emergence of features of a number of skin pathologies of people living near uranium tailings in a mountain ecosystem, we assume that the endogenous intake and gradual removal of radionuclides from the body does not lead to a rapid onset of effect, but over time causes a number of epigenetic effects that manifest themselves phenotypically. Our assumptions are confirmed by a number of works in the field of environmental epigenetics [44–47].

A significant role in the development of the "effect" is played by the characteristics of the barometric pressure of the atmosphere in the conditions of a mountain ecosystem. Thus, the combined effect of ionizing radiation and natural hypoxia in a mountain ecosystem modifies the radiosensitivity and radioresistance of the body, changing oxidative metabolism. When studying this issue, it is necessary to take into account a number of provisions of classical radiobiology, one of which is the "oxygen effect" [48]. This effect has been known to science for a long time, but recently it has been studied in more detail not for protection, but, on the contrary, for a more effective effect on tumor cells during treatment with ionizing radiation at high partial pressure of oxygen [48].

Thus, it is important to take into account the modifying effect of hypoxia in mountainous areas when developing algorithms for carrying out therapeutic and preventive measures among people exposed to endogenous exposure to low doses of radiation.

The identified severe deficiency of vitamin D in people living in geochemical provinces in the mountain ecosystem maintains the pathological condition of the skin. The hormonal form of vitamin D, in addition to regulating calcium homeostasis, has important pleiotropic effects mediated through the vitamin D receptor (VDR) and affects almost all body functions [49]. Its regulatory ability influences the proliferation and differentiation of keratinocytes [50], as well as innate and adaptive immunity [51], thereby helping to restore the epidermal barrier and protective properties of the skin.

The next important component of the skin that we studied in our work is the skin microbiome. The health of the human integumentary system largely depends on its quantitative and qualitative characteristics [52]. It should be noted that the skin microbiome is very sensitive to changes in environmental factors. Its changes lead to damage to the skin. But, on the other hand, thanks to increased adaptive abilities, the skin microbiota is able to positively influence the immunological mechanisms in the skin and have a regulatory role aimed at the protective properties of the skin under normal conditions [52, 53].

Under conditions of technogenic pollution, a damaged epidermal barrier cannot be a favorable habitat for saprophytic microbiota. Such conditions close the "pathological circle" by promoting the colonization of the skin by pathogenic flora. Thus, in our study, we prove that damage to the epidermal barrier under conditions of technogenic pollution determines the characteristics of the manifestation of dermatoses, their course, frequency of occurrence, and the presence of comorbidity. At the same time, depletion of the skin's reserve capabilities and disruption of its regenerative abilities lead to premature aging of the skin and the development of skin tumors.

When creating an algorithm for the treatment of skin diseases, it is necessary to take into account the influence of technogenic impacts and the specific living conditions in the mountain ecosystem of Kyrgyzstan.

3. Conclusions

Analysis of the structure, prevalence, clinical and anamnestic features of the course, and manifestations of dermatoses showed a high frequency of infectious dermatoses with a deeper protracted course, especially in former miners of uranium enterprises. The current situation is a reflection of immunodeficiency states, as indicated by our immunological studies, where the inhibition of the activation function of T- and B-lymphocytes clearly correlates with a decrease in the production of immunoglobulins, as well as with a decrease in the number and functional activity of mononuclear phagocytes and neutrophils (**Figure 11**, **Table 2**). This situation is exacerbated by a change in the microbiological profile of the skin (**Table 1**) toward the predominance of pathogenic flora (**Figure 10**). The average age of manifestation of gerantogenesis stigmas (steatosis, keratoderma of the palms and soles, dyschromia, diffuse hair thinning, early graying, senile lentigo, seborrheic keratosis, senile angiomas) was 32.7 ± 7.5 years compared with the control group 45,2 ± 10 years. A relationship was found between the risk factor and the frequency of occurrence of the main age-related stigmas in people living near uranium tailings in mountainous conditions. The influence of risk factors on the likelihood of age-related stigma from 37% ± 2.82% (95% CI 31.5% - 42.5%) to 63.2% ± 2.81 (95% CI 57.7% - 68, 7%), in addition to the probability that exists for persons who have not been exposed to a negative factor. Violations of the epidermal barrier, a decrease in skin resistance to environmental factors, and the presence of dystrophic changes in the skin appendages suggest structural and functional changes in the human integumentary system as a result of a long-term, predominantly endogenous, intake of radionuclides into the internal environment of the body.

Comorbidity in dermatological pathology is associated, on the one hand, with a violation of the skin barrier, and on the other hand, with changes in the main links of cellular and humoral immunity toward immunosuppression.

Based on the data obtained, it is necessary to develop therapeutic and preventive measures to provide practical assistance to residents of uranium geochemical provinces, which can be a model in the event of a radiation hazard.

Author details

Alena Isupova
Kyrgyz-Russian Slavic University, Kyrgyz State Medical Academy, Bishkek, Kyrgyzstan

*Address all correspondence to: alena.isupova.2015@mail.ru

References

[1] Lespukh E, Stegnar P, Usubalieva A, Solomatina A, Tolongutov B, Beishenkulova R. Assessment of the radiological impact of gamma and radon dose rates at former U mining sites in Kyrgyzstan. Journal of Environmental Radioactivity. 2013;**123**:28-36. DOI: 10.1016/j.jenvrad.2012.11.013

[2] Latifi-Pupovci H, Selmonaj M, Ahmetaj-Shala B, Dushi M, Grajqevci V. Incidence of haematological malignancies in Kosovo-a post "uranium war" concern. PLoS One. 2020;**15**(5):e0232063. DOI: 10.1371/journal.pone.0232063

[3] Berdibaeva AB, Bykovchenko YuG, Zhunushov AT, Tukhvatshin RR. Impact of Uranium Tailings on Biospheric Objects in Kyrgyzstan: Medicine of Kyrgyzstan. 2014. No. 2. Available from: https://cyberleninka.ru/article/n/vozdeystvie-uranovyh-hvostohranilisch-na-biosfernye-obekty-v-kyrgyzstane [Accessed: October 07, 2023]. (In Russian)

[4] Kaldybaev B, Chinara SK. Radioecological researches in technogenic areas of Issyk-Kul region. KnE Energy. 2018, 2018;**3**(2):129-133. DOI: 10.18502/ken.v3i2.180

[5] Bykovchenko Yu G, Bykova EI, Belekov T, Tukhvatshin RR. Technogenic Pollution by Uranium of the Biosphere of Kyrgyzstan. Bishkek: AO Altyn-Tamga, 2005; 186 p. (In Russian)

[6] Climate Change Adaptation and Mitigation in the Kyrgyz Republic. IMF Staff Country Reports. 2023;**092**:A002. DOI: 10.5089/9798400232725.002.A002

[7] Corcho Alvarado JA, Balsiger B, Röllin S, Jakob A, Burger M. Radioactive and chemical contamination of the water resources in the former uranium mining and milling sites of Mailuu Suu (Kyrgyzstan). Journal of Environmental Radioactivity. 2014;**138**:1-10. DOI: 10.1016/j.jenvrad.2014.07.018

[8] Vandenhove H, Sweeck L, Mallants D, Vanmarcke H, Aitkulov A, Sadyrov O, et al. Assessment of radiation exposure in the uranium mining and milling area of Mailuu Suu, Kyrgyzstan. Journal of Environmental Radioactivity. 2006;**88**(2):118-139. DOI: 10.1016/j.jenvrad.2006.01.008

[9] Lind OC, Stegnar P, Tolongutov B, Rosseland BO, Strømman G, Uralbekov B, et al. Environmental impact assessment of radionuclide and metal contamination at the former U site at Kadji Sai, Kyrgyzstan. Journal of Environmental Radioactivity. 2013;**123**:37-49. DOI: 10.1016/j.jenvrad.2012.07.010

[10] Salbu B, Burkitbaev M, Strømman G, Shishkov I, Kayukov P, Uralbekov B, et al. Environmental impact assessment of radionuclides and trace elements at the Kurday U mining site, Kazakhstan. Journal of Environmental Radioactivity. 2013;**123**:14-27. DOI: 10.1016/j.jenvrad.2012.05.001

[11] Skipperud L, Strømman G, Yunusov M, Stegnar P, Uralbekov B, Tilloboev H, et al. Environmental impact assessment of radionuclide and metal contamination at the former U sites Taboshar and Digmai, Tajikistan. Journal of Environmental Radioactivity. 2013;**123**:50-62. DOI: 10.1016/j.jenvrad.2012.05.007

[12] Uranium Tailings in Central Asia: The Case of the Kyrgyz. REPORTOct 15, 2009 Republichttps. Available from: www.nti.org/analysis/articles/uranium-tailings-kyrgyz-republic/

[13] Leng P, Zhang Q, Li F, Kulmatov R, Wang G, Qiao Y, et al. Agricultural impacts drive longitudinal variations of riverine water quality of the Aral Sea basin (Amu Darya and Syr Darya Rivers), Central Asia. Environmental Pollution. 2021;**284**(1):117405. DOI: 10.1016/j.envpol.2021.117405

[14] Hao L, Wang P, Gojenko B, Yu J, Lv A, Li F, Kenjabaev S, Kulmatov R, Khikmatov F. Five decades of freshwater salinization in the Amu Darya River basin. Journal of Hydrology: Regional Studies. 2023;**47**:101375. DOI:10.1016/j.ejrh.2023.101375

[15] Zhang Q, Chen Y, Li Z, Xiang Y, Li Y, Sun C. Recent changes in glaciers in the northern Tien Shan, Central Asia. Remote Sensing. 2022;**14**(12):2878. DOI: 10.3390/rs14122878

[16] Barandun M, Pohl E, Naegeli K, McNabb R, Huss M, Berthier E, et al. Hot spots of glacier mass balance variability in Central Asia. Geophysical Research Letters. 2021;**48**(11):e2020GL092084. DOI: 10.1029/2020GL092084

[17] Hao L, Wang P, Yu J, Ruan H. An integrative analytical framework of water-energy-food security for sustainable development at the country scale: A case study of five central Asian countries. Journal of Hydrology. 2022;**607**:127530. DOI: 10.1016/j.jhydrol.2022.127530

[18] Delbart W, Karabet J, Marin G, Penninckx S, Derrien J, Ghanem GE, et al. Understanding the radiobiological mechanisms induced by 177Lu-DOTATATE in comparison to external beam radiation therapy. International Journal of Molecular Sciences. 2022;**23**:12369. DOI: 10.3390/ijms232012369

[19] Li Y, Chyan MK, Hamlin DK, Nguyen H, Corey E, Wilbur DS. Oxidation of p-[125I]Iodobenzoic acid and p-[211At]Astatobenzoic acid derivatives and evaluation In vivo. International Journal of Molecular Sciences. 2022;**23**:10655. DOI: 10.3390/ijms231810655

[20] van der Wal BCH, Dadachova E. Targeted radionuclide therapy of cancer and infections. International Journal of Molecular Sciences. 2023;**24**(10):9081. DOI: 10.3390/ijms2410908121

[21] Idrissou MB, Pichard A, Tee B, Kibedi T, Poty S, Pouget JP. Targeted radionuclide therapy using auger electron emitters: The quest for the right vector and the right radionuclide. Pharmaceutics. 2021;**13**(7):980. DOI: 10.3390/pharmaceutics13070980

[22] Korde A, Mikolajczak R, Kolenc P, et al. Practical considerations for navigating the regulatory landscape of non-clinical studies for clinical translation of radiopharmaceuticals. EJNMMI Radiopharmacy and Chemistry. 2022;**7**:18. DOI: 10.1186/s41181-022-00168-x

[23] Lourenço J, Mendo S, Pereira R. Radioactively contaminated areas: Bioindicator species and biomarkers of effect in an early warning scheme for a preliminary risk assessment. Journal of Hazardous Materials. 2016;**5**(317):503-542. DOI: 10.1016/j.jhazmat.2016.06.020

[24] Shaki F, Zamani E, Arjmand A, Pourahmad J. A Review on Toxicodynamics of Depleted Uranium. Iranian Journal of Pharmaceutical Research. 2019;**18**:90-100. DOI: 10.22037/ijpr.2020.113045.14085

[25] Guéguen Y, Frerejacques M. Review of knowledge of uranium-induced kidney toxicity for the development of an adverse outcome pathway to renal

impairment. International Journal of Molecular Sciences. 2022;**23**(8):4397. DOI: 10.3390/ijms23084397

[26] Dinocourt C, Culeux C, Legrand M, Elie C, Lestaevel P. Chronic exposure to uranium from gestation: Effects on behavior and neurogenesis in adulthood. International Journal of Environmental Research and Public Health. 2017;**14**(5):536. DOI: 10.3390/ijerph14050536

[27] Bjorklund G, Semenova Y, Pivina L, Dadar M, Rahman MM, Aaseth J, et al. Uranium in drinking water: A public health threat. Archives of Toxicology. 2020;**94**(5):1551-1560. DOI: 10.1007/s00204-020-02676-8

[28] Kelly-Reif K, Sandler DP, Shore D, Schubauer-Berigan M, Troester MA, Nylander-French L, et al. Mortality and cancer incidence among underground uranium miners in the Czech Republic 1977-1992. Occupational and Environmental Medicine. 2019;**76**(8):511-518. DOI: 10.1136/oemed-2018-105562

[29] Asic A, Kurtovic-Kozaric A, Besic L, Mehinovic L, Hasic A, Kozaric M, et al. Chemical toxicity and radioactivity of depleted uranium: The evidence from in vivo and in vitro studies. Environmental Research. 2017;**156**:665-673. DOI: 10.1016/j.envres.2017.04.032

[30] Birchall A, Puncher M, Marsh JW, Davis K, Bailey MR, Jarvis NS, et al. IMBA professional plus: A flexible approach to internal dosimetry. Radiation Protection Dosimetry. 2007;**125**(1-4):194-197. DOI: 10.1093/rpd/ncl171

[31] ICRP Publication 107. Nuclear decay data for dosimetric calculations. Annals of the ICRP. 2008;**38**(3):7-96. DOI: 10.1177/01466453211016658

[32] Kai M. ICRP recommendations for recovery. Annals of the ICRP. 2021;**50**:8-14. DOI: 10.1177/01466453211006809

[33] Rump A, Becker B, Eder S, Lamkowski A, Abend M, Port M. Medical management of victims contaminated with radionuclides after a "dirty bomb" attack. Military Medical Research. 2018;**5**(1):27. DOI: 10.1186/s40779-018-0174-5

[34] Keith S, Faroon O, Roney N, et al. Toxicological Profile for Uranium. Atlanta (GA): Agency for Toxic Substances and Disease Registry (US); 2013. p. 2, Relevance to Public Health. Available from: https://www.ncbi.nlm.nih.gov/books/NBK158801/

[35] Algrifi MA, Salman TM. Measuring uranium concentration in Um Qasr district, southern Iraq, in two different ways. Applied Radiation and Isotopes. 2023;**192**:110595. DOI: 10.1016/j.apradiso.2022.110595

[36] Ma M, Wang R, Xu L, Xu M, Liu S. Emerging health risks and underlying toxicological mechanisms of uranium contamination: Lessons from the past two decades. Environment International. 2020;**145**:106107. DOI: 10.1016/j.envint.2020.106107

[37] National Statistical Committee of the Kyrgyz Republic. 2023. Available from: http://stat.kg/en/

[38] Transboundary Monitoring and Information System for Radioactive Contamination in the Event of Natural Hazards. 2020. Available from: https://casib.eu/en/node/42

[39] Friedman RJ, Rigel DS, Kopf AW. Early detection of malignant melanoma: The role of physician examination and self-examination of the skin.

CA: a Cancer Journal for Clinicians. 1985;**35**(3):130-151 DOI:10.3322

[40] Steiner A, Pehamberger H, Wolff K. In vivo epiluminescence microscopy of pigmented skin lesions. II. Diagnosis of small pigmented skin lesions and early detection of malignant melanoma. Journal of the American Academy of Dermatology. 1987;**17**(4):584-591. DOI: 10.1016/s0190-9622(87)70240-0

[41] Isupova A.A. "The Experience of Using the Method of Dermatoscopy in a Comprehensive Study of the Skin of People Long-Term Living in the Area of Industrial Pollution" [thesis]. Material of the 24th World Congress of Dermatology, Milan, Italy, 2019 (WCD2019). Available from: https://www.wcd2019milan.org/ l

[42] Isupova AA. Features of age-related changes in the skin in persons living near uranium tailings in the mountains. Radiation and Risk. 2022;**31**(1):29-39. DOI: 10.21870/0131-3878-2022-31-1-29-39 (In Russian)

[43] Isupova AA, Isupov RV. Frequency of detection of reference values of vitamin D levels in the blood serum of the indigenous inhabitants of Kyrgyzstan. Profilakticheskaya Meditsina. 2022;**25**(4):55-60. DOI: 10.17116/profmed20222504155 (In Russian)

[44] Bernal AJ, Jirtle RL. Epigenomic disruption: The effects of early developmental exposures. Birth Defects Research. Part A, Clinical and Molecular Teratology. 2010;**88**(10):938-944. DOI: 10.1002/bdra.20685

[45] Kolaitis ND, Finger BJ, Merriner DJ, Nguyen J, Houston BJ, O'Bryan MK, et al. Impact of chronic multi-generational exposure to an environmentally relevant atrazine concentration on testicular development and function in mice. Cells. 2023;**12**(4):648. DOI: 10.3390/cells12040648

[46] Richards CL, Alonso C, Becker C, Bossdorf O, Bucher E, Colomé-Tatché M, et al. Ecological plant epigenetics: Evidence from model and non-model species, and the way forward. Ecology Letters. 2017;**20**(12):1576-1590. DOI: 10.1111/ele.12858

[47] Santaló J, Berdasco M. Ethical implications of epigenetics in the era of personalized medicine. Clinical Epigenetics. 2022;**14**(1):44. DOI: 10.1186/s13148-022-01263-1

[48] Perillo B, Di Donato M, Pesone A, Di Zazzo E, Giovannelli P, Galasso G, et al. ROS in cancer therapy: The bright side of the moon. Experimental & Molecular Medicine. 2020;**52**(2):192-203. DOI: 10.1038/s12276-020-0384-2

[49] Rochel N. Vitamin D and Its Receptor from a Structural Perspective. Nutrients. 2022;**14**(14):2847. DOI: 10.3390/nu14142847

[50] Zhang J, Zhu Q, Zhang S, Wu J, Huang J, Li M, et al. Double knockout of vitamin D receptor and its coactivator mediator complex subunit 1 unexpectedly enhances epidermal permeability barrier function in mice. Biochimica et Biophysica Acta, Molecular Cell Research. 1868;**2021**(12):119131. DOI: 10.1016/j.bbamcr.2021.119131

[51] Hewison M. Vitamin D and innate and adaptive immunity. Vitamins and Hormones. 2011;**86**:23-62. DOI: 10.1016/B978-0-12-386960-9.00002-2

[52] Grice EA, Segre JA. The skin microbiome. [published correction appears] Nature Reviews. Microbiology. 2011;**9**(4):244-253. DOI: 10.1038/nrmicro2537

[53] Skowron K, Bauza-Kaszewska J, Kraszewska Z, et al. Human skin microbiome: Impact of intrinsic and extrinsic factors on skin microbiota. Microorganisms. 2021;**9**(3):543. DOI: 10.3390/microorganisms9030543